A Boy on the Farm

Joseph E. Narigon

Hayseed Press LLC

Published by Hayseed Press LLC
hayseedpress.com
Cedar Falls, Iowa

ISBN 979-8-9928776-1-8

Cover Design by Aaron and Nick Narigon
Cover photo of Joseph E. Narigon and Effie Lee (Narigon) Boggess

FOREWORD

The Witness Tree. Photo by David O. Narigon

"I have lived a long time. In my many years, I have seen a lot of changes, both good and bad, but I didn't like a damn one of them." My grandfather Virgil Narigon made this comment to me during our last conversation a few days before his death at the age of 93.

Uncle Joe's memoir, *A Boy on the Farm*, is a collection of stories that harken back to simpler times when farmers like Virgil relied on manpower and horse power. Part of their manpower was their children who grew up knowing they were expected to lend a hand. They took on chores at an early age: gathering eggs, rounding up the milk cows and driving them to the barn, weeding, and harvesting the garden.

When Dad yelled you were expected to come running to do whatever task he was wanting. As you got older the tasks changed with your experience and strength: picking up hay, building fence, etc. There was no sexual discrimination. Daughters were expected to work along with brothers, sisters, fathers, and mothers. All had chores to be completed after school and before dinner. Children in an ag economy were assets not expense. Uncle Joe, his sister Effie Lee, his brother Don (my dad), and his baby sister Eleanor grew up with these expectations.

Uncle Joe's life witnessed the transformation of farming from primarily manual—working side-by-side with teams of horses—to primarily mechanized. It would take months to pick corn by hand. Now modern combines can pick that same amount of corn in a few days. The size of farms were limited by how much work you could get done with your family and hired hands. Most farmers had at least one hired hand that lived on the property in one of the small houses that had probably been built in the in the late-19th or early-20th century—when the average farm was 40 acres.

By 1940 the average farm in the U.S. was 178 acres and in 2020 it was 448, so a little over an entire family was eliminated from farming. As farms increased in size they needed to purchase bigger equipment to get things done on a timely basis. Farming is a race with mother nature to find the weather necessary to plant, to hay, and harvest. You no longer have time to cultivate your row crops and every six weeks remove weeds. There's no longer time to hire kids to weed your beans or pick up your hay.

In the '60s and '70s agriculture became increasingly dependent on chemicals and fertilizers. As grandfather said, not all changes are good. With the need to increase yields in a timely basis we polluted our streams and groundwater with chemicals. Erosion increased as well. The transformational change in agriculture also resulted in a transformational change in our small rural communities where service jobs relied on all those farm families that were no longer there. There was reduced needs for appliance, hardware, clothing, and car dealers.

Also our schools had declining enrollments, which dramatically reduced budgets. To accommodate the farmers we built infrastructure in the way of roads and bridges. At one time almost every square mile was bordered by either a dirt or gravel road along with the necessary bridges. Now and in the future there looms a constant battle between urban and rural as to how much tax dollars can be targeted to these infrastructural needs when rural areas generate less taxes to the general fund. There will be continued pressure to consolidate schools and county government, close roads and not maintain bridges. Does Iowa continue to need 99 counties?

Uncle Joe's stories are not only about the people but the land. There is a bur oak in the yard northwest of Uncle Joe's boyhood home where I live now. This tree is now named the Witness Tree (pictured). At four-and-half feet above the ground the circumference of the tree is 105-inches. With a growth factor of five, the estimated age of that tree is 525 years old. So it could have started growing about the time Columbus took his voyage to the New World. In its life it witnessed a lot of changes. Some good and some bad—but from whom's perspective.

The hope for the future for the house and acreage, along with the 177 acres of the homestead farm of great-great grandfather Joseph Narigon whom Uncle Joe was named after, is that it will all remain in the family. All the properties are free of debt. Hopefully family and friends can enjoy the properties for the next 150 years and beyond. I am continuing to rehab the house, which will be completed and perhaps opened as a bed and breakfast. We might also have spots available for camping and RVs. So come sit a spell and rock and we can share stories.

Blessings everyone.

—David O. Narigon

INTRODUCTION

This is a story about my life growing up on a farm in Adams County, Iowa. Adams County is located in southwest Iowa approximately 75 miles southeast of Omaha, Nebraska. The main highway running through the county is US Highway 34. The county is composed of 425 square miles divided into twelve townships. It is a rural county that at one time had a population of over 13,500. The average farm size then was around 40 acres and most of the population lived on farms. As the farm size increased the population decreased. Today (2023) the population is less than 4,500, making Adams the least populous county in Iowa. This information is provided to establish the setting for my story.

The purpose of my story is to provide a history of farm life as I experienced it so my children, grandchildren, and great-grandchildren will know more about their family heritage. I hope they find it entertaining. Information was gathered from the Adams County History published in 1984; *Nodaway Iowa - Past and Present*, published in 1976, and *The Family History of My Grandparents*, compiled by Mark Boggess as a class assignment in high school. Mark is the son of my sister, Effie Lee.

Most of the stories are from my memory and may or may not be entirely accurate. Additional information on the invention and history of farm machines was gathered using the internet. Pictures are from my boyhood album and a replica of the 1908 Sears & Roebuck catalogue.

Thanks to Ed Narigon, Nicholas Narigon, and Betty Narigon for their suggestions and help in editing the story, and thanks to Nancy Nass of Holiday Island, Arkansas, for her computer advice.

—*Joseph E. Narigon, 2003*

[Editor's note: Joseph E. Narigon originally self-published this memoir in 2003. Following his passing in 2023, Joe's family updated and republished the memoir to share his legacy with future generations.]

1

FAMILY

This picture of Dad, Mother, Effie Lee, and me was taken
when I was one month old

March 27, 1929

It was a cold and cloudy day. (I always liked, "Dark and Stormy Night," the way Snoopy started his books.) But this event happened in daytime in a small farmhouse about two or three miles northwest of Brooks, Iowa. Brooks was a small town in Adams County, five miles southwest of Corning. Corning, the county seat, was where the closest doctor office was located.

Sometime in the early morning of March 27, 1929, a doctor from Corning was summoned to the home of Virgil Narigon to deliver a baby. I don't know if the Narigons had a phone or if a neighbor called, but the doctor arrived. He may have ridden a horse, or traveled by horse and buggy if the roads were muddy. In those years, all babies were delivered at home, unless there were severe complications.

A baby boy was born and that baby boy, Joseph Edward Narigon, was me. I was the second child of Virgil Oscar Narigon and Flossie (Brown) Narigon. My sister, Effie Lee, was born on December 27, 1927. The family grew more when my brother Don was born on November 16, 1934. Elinor, the last child in the family, was born August 19, 1936. They were also delivered at home. My fate as the middle child was sealed, and I have fit into that mold all my life.

Me (left) and Effie Lee

My parents decided to use names of their ancestors to name their children. I was the namesake of my great-grandfather Joseph (1840-1908). Joseph was the founder of the Narigon family in Iowa and the only family group with the last name spelled Narigon. He was born in 1840 in Tuscarawas County, Ohio, the son of Nicholas and Mary Narragong.

Throughout the generations there has been some confusion over the spelling of the last name Narigon. Historical records spell the name from "Narrgang" to "Naragong" and everything in between. Nicholas was born in 1800 in Virginia to Jacob Naragong, or Norgang, and Maria Barbara Gottschall. Jacob was born in 1773 in Montgomery, Pennsylvania. His father, Daniel Nargang II was born in 1738 in Berks County, Pennsylvania, British Colonial America.

His father, Daniel Nargang, was the first Narigon to sail to America. Daniel Nargang was born in 1715 in Leistadt, Germany, which was then part of the Holy Roman Empire. His parents were Johannes and Anna Elisabeth Naergang. During the Nine Years' War in the late 1600s, Leistadt, and the whole state of Rhineland-Palatinate, was attacked by the French. Subsequent battles continued into the early 18th century. William Penn visited the area as early as 1677 to recruit Germans to relocate to his new colony in America.

Thousands of German refugees, known as Palatines, fled first to England where Queen Anne made them "denizens of the kingdom without charge." From there they made their way to British Colonial America where free passage was given and free land awaited. The first Palatine ship set sail in 1708.

Daniel Nargang and his two older brothers sailed from Rotterdam to Philadelphia on the Ship Harle in 1736. (Also on the ship was the ancestor of the founder of the Studebaker Motor Car Company.) The Nargangs settled in Berks County, Pennsylvania. There Daniel Nargang married a woman named Eva. Daniel Nargang II was the first of their two recorded children.

Jacob Naragong was the sixth child of Daniel Nargang II and Susanna Catherine Smith. Jacob married Maria Barbara Gottschall in 1798. Following the Treaty of Greenville in 1795, Moravian missionaries established a mission in Tuscarawas County in western Ohio. Farmers from Pennsylvania followed soon after to settle the land.

Around 1802, between the birth of their second and third child, Jacob and Maria moved to Jefferson Township in Tuscarawas County to farm. They had a total of 11 children. Jacob fought in the War of 1812. The family legend is that Jacob served with Commodore Oliver Hazard Perry in the Battle of Lake Erie. Jacob suffered many hardships during the war. His shoes wore out and his feet were so severely frozen that he lost his toes. For his valiant service, Jacob was deeded a tract of land in Ohio.

Jacob Naragong's gravestone in Blue Ridge, Harrison County, Ohio says he died August 2, 1846, age 73 years

When Jacob's oldest son Nicholas married Mary Wilson in 1832, the last name on Nicholas' marriage certificate was spelled "Narragong." On his death record from 1876, Nicholas' last name was spelled "Norigon." Nicholas and Mary had six children, all born in Harrison County, Ohio. William was their fourth and Joseph, three years younger, was their fifth.

In 1861, my great-grandfather Joseph Narragan (as it was spelled on his enlistment papers) and his older brother, William, enlisted in Company E. of the 80th Ohio Infantry to fight for the Union side in the Civil War. They fought in the battles of Iuka, Corinth, Raymond, Jackson, and the siege of Vicksburg.

On November 25, 1863, at the battle of Missionary Ridge in Chattanooga, Tennessee, under the command of General William Tecumseh Sherman, Joseph was severely injured. William was killed in battle, and it is said Joseph was shot in the act of lifting his dying brother. The ball entered Joseph's right side, breaking three ribs and his shoulder blade.

Joseph was taken to the field hospital on the Tennessee River, where he remained for two weeks before transferring to the hospital at Chattanooga. At some point one bone was removed from his arm. In Chattanooga, Joseph lay on his back for nine months. He finally made it home to Ohio and was honorably discharged on September 22, 1864.

After the war, Joseph married Adelaide Humphrey, in Bushnell, Illinois. Joseph and Adelaide's first two children were born in Pella, Iowa, and they moved to Adams County in 1872 in time for the birth of their third child, William Tecumseh. My grandfather, Jacob Oscar, was their fourth child. By this time the family changed the spelling of their last name to "Narigon." So if you meet anyone with the last name, Narigon, you know they are a direct descendant of my great-grandfather Joseph.

Joseph died suddenly of heart failure at age 67 at the family farm in Nodaway in 1908. His obituary says, "He was scrupulously honest in his dealings and a sturdy supporter of what he thought was right, a good neighbor, a faithful friend, and a good husband and father."

From left, me, Don, and Effie Lee

I was named after my great-grandfather Joseph Narigon. My sister, Effie Lee, was named for our grandmother, Effie Narigon. Don Oscar was named after our grandfather, Jacob Oscar Narigon, and Elinor Genevieve received her name from our grandmother, Genevieve Brown. I do recall it took several weeks or months to decide what to call Elinor, and she went by "It" until she was officially named. In fact Dad called her "It" most of her life.

I do not remember living in the house where I was born, as we moved in 1932 when I was three. I do remember visiting the house several years later. It was a small one-story house with no electricity or running water. It was heated with a wood stove, and I am sure there were many cold mornings in the winter. The house and other farm buildings have been gone for years.

My Parents

My mother and father on their wedding day in 1927

My father, Virgil Oscar Narigon, was born June 8, 1903, at the home of Oscar and Effie Narigon near Highland in Adams County, Iowa. He was the oldest of three children. Ethel was born November 6, 1906, followed by John on May 9, 1908. Virgil's mother, Effie, died September 13, 1914, when he was 11 years old.

Dad graduated from the Corning High School in 1922. His claim to fame was running the mile relay on the track team. The mile relay team recorded the second fastest time in the state his senior year. He kept his track shoes with him until the day he died.

After graduation Dad worked for local farmers and spent several years following the wheat harvest from Iowa to North Dakota. Dad and Mother were married in Omaha, Nebraska on January 27, 1927, and started farming on the place where Effie Lee and I were born northwest of Brooks, Iowa.

Dad was a disciplinarian and a man of few words. When he gave an order, he expected immediate action. He shaved with a straight edge razor and kept it sharp with a leather razor strap. The strap had two straps layered together and when swung, the straps popped together making a loud sound. I can remember when Effie Lee and I were fighting over the hot air register getting dressed for school he would get upset and come at us with the razor strap. I would quickly sit in a little chair and he would hit the back of the chair making lots of noise. I would cry like I was really getting hurt. He never hit us hard but made believers of us to do what he said when he said it. I think he mellowed in later years as I don't remember Don or Ellie ever getting disciplined. Maybe they were never in trouble.

Dad was a hard worker and very smart in figuring out ways to make money. Going through the Great Depression in the 1930s made a lasting impression on him and the way he spent his money. He really enjoyed his grandchildren, all 18 of them. He would tease the girls on how their boyfriends looked like gerbils. He would catch the little ones as they ran by his chair with his cane, steal their noses, and threaten them that if they weren't good he would put them in the spanking machine that was down in the basement.

Dad and Mother in 1980

Dad lived a long and relatively happy life. We had great celebrations for his 80th and 90th birthdays with nearly all his children and grandchildren attending. He was lonely after Mother died but was able to stay on the farm until his last few years when he stayed in a nursing home in Corning. He died on July 27, 1995, at the age of 92 years, one month, and 19 days.

Mother was born on April 17, 1902, to Ezra C. and Genevieve Day Brown. She was named Flossie. At least she thought that was her name until in later life she got a copy of her birth certificate and found her name was recorded as Florence Eloise. She was unhappy about her legal name and always went by Flossie. She was the oldest of six children and was always the peace maker in the family.

Mother was born on a farm in Page County. She attended a rural school near her home through the eighth grade. She then stayed in Villisca with her grandparents, the Edward Days, and graduated from Villisca High School. After high school she taught in rural schools until her marriage to Dad.

Flossie was a great mother. I can remember how she read to us when we were little and always tried to solve our problems. She was a wonderful cook and kept a very clean house.

There was always love expressed in our home but it was never vocal. The only time I ever heard Dad tell Mother he loved her was at the hospital after she had died on August 19, 1983.

Effie May and Oscar Narigon

Grandparents and Ancestors

I was limited in the number of grandparents I knew as I grew up. My father's mother, Effie May Bixler Narigon, died before I was born. The youngest of 13 children, Effie May was born in 1873 on a farm nearby in Brooks. Her father, John Bixler, Sr., was born in Washington County, Pennsylvania in 1823. His father, Jacob Bixler, was born in 1796 in Fayette County, Pennsylvania, the youngest son of Joseph Bixler (1755-1805) and Elizabeth Susanna Strickler (1757-1830). Joseph Bixler was a prominent miller in York, Pennsylvania. His parents, Christen Bicksel, Sr., and Catherine Shearer were from Bern, Switzerland and immigrated to British Colonial American in 1749.

Jacob and Elizabeth Murray Bixler (whose father Robert Murray emigrated from Ireland) moved to Ohio when John Bixler, Sr. was eight years old. John Senior married Savilla Ann Markley, there in Ohio in 1846. In October 1854 John Sr. and Savilla moved 600 miles to Adams County with two teams of oxen, their household building materials and utensils, and 75 cents in their pocket. John Sr.'s parents, Jacob and Elizabeth, followed them to Adams County two years later. (In 1862, at the beginning of the Civil War, Jacob entered military service at age of 62.)

At first it was tough going for the Bixlers in Adams County. According to my Aunt Louise, a prairie fire in the fall of 1856 destroyed feed, grass, chickens, outhouses, and stables. That winter of 1856-1857 saw heavy snowstorms. In some places the snow drifts were thirty- to fifty-feet-deep. You couldn't tell the hills from the hollows. Herds of deer would fall through the crust and drown in the snow, or get killed by wolves or hunters. Due to the late frost no corn matured in the summer of 1857 and they had to haul corn for bread and seed from Missouri.

Still, John and Savilla, who were some of the earliest settlers of Adams County, saw the county go from wilderness to prosperous farmland. They had thirteen children, eleven of whom survived to adulthood. Effie May was the youngest. She took care of her parents after John Sr. suffered paralysis in 1900 and until his death in 1903.

From left, back row, Effie May and Oscar Narigon; front row, my aunt Ethel and uncle John in 1912

Effie May was only 41 at the time of her death in 1841. Dad was only 11, Aunt Ethel was eight, and Uncle John was six. Granddad never remarried and raised the three children by himself with the help of a housekeeper.

I do remember Grandfather Narigon (Jacob Oscar—who went by Oscar). My uncle John lived with him and they farmed together at the north place where Dad was born. Dad's sister, Ethel, was away from home teaching school but came home in the summer and on school holidays. We went to Granddad's place for Sunday dinner and other parties.

Grandfather Narigon was one of eight children, four boys and four girls. One of the boys, David, died before I was born. The other two brothers, John and Bill, lived in the Nodaway area as did two of the sisters. We went to see them at different times during the year. About twice a year all the Narigons got together and you could depend on some big argument getting started. It got louder and louder but I cannot remember any physical altercations taking place.

Grandfather Narigon died in 1934 and I remember going to his funeral. I think it was the first funeral that I attended. The church was full. We were driven to the cemetery and back to his house for lunch. The funeral home provided cars and drivers, and I can remember riding in the middle of the front seat. All cars were straight stick shifts, and I was so interested in how the man driving the car drove over a mile without touching any of the pedals on the floor of the car. There is a small slope going out the cemetery and I think he coasted down the hill. We were going so slow he didn't need to push the gas pedal.

John continued to live on and farm the north place until he got married. In 1939, Dad purchased the farm from John and Ethel. Dad rented out the farmhouse and Ethel lived with us when she came home from teaching. The schools where she taught had electricity and had decorations for the Christmas trees. She brought all the decorations for our tree, but I was always disappointed because we could not light the bulbs.

Uncle John and Dad had a disagreement on the sale of the farm, and we didn't see him very often after John was married and moved to a farm near Prescott. Ethel continued to spend her summers and holidays with us. In the summer I wore my overalls without a shirt and really got brown from the sun. Ethel always thought I looked dirty and scrubbed me in the tub trying to get all the dirt off.

Grandfather (Ezra) Brown, my mother's father, died in an accident in December of 1929. He was killed in a cornfield by an enraged Jersey bull. He went out to the field in the morning and was missing for a few hours. A search party found his body trampled with a broken club nearby. Also nearby was a three-year-old bull who was known to have a vicious temper. I was only nine months old so do not remember him at all.

Ezra's grandfather, Spencer Hadley Brown, was a pioneer doctor in Missouri who served as a physician for the Union forces during the Civil War. Spencer's grandfather, Jacob Brown, was born to a Quaker family in Guilford County, North Carolina in 1755. His wife Mary was from Bucks County, Pennsylvania, and they were disowned by the Quakers in 1776 for "marrying out of unity," which means the married couple were of different religions. Jacob was further chastised by the Quakers for taking up arms to fight in the Revolutionary War.

Jacob's father, Thomas Brown IV, was born in Bucks County, Pennsylvania in 1728. His father, Thomas Brown III, was born in Argyll and Bute, Scotland in 1691. He came to America by himself at age 15 in 1706, likely as an indentured servant. He married Ruth Large and had eight children.

(Ruth's great-grandfather William Large emigrated with his family in 1635 from Hingham, England to New Hingham in Plymouth County, Massachusetts Bay Colony. Abraham Lincoln's ancestor Samuel Lincoln immigrated to New Hingham two years later in 1637.)

In 1740, Thomas, Ruth, and seven of their eight children migrated along with other Quakers to Shenandoah Valley, Virginia. Thomas Brown III built a log cabin in Inwood, West Virginia (then Virginia). He was one of the first to grow a fruit orchard in the area. He assembled a total of 1,200 acres of land. When he died in 1750 his will divided the land up amongst Ruth and all of his children. Thomas Brown IV received 200 acres in North Carolina. Ruth received the original log cabin and 60 acres. The log cabin is still there. The Thomas Brown House is the oldest house in Berkeley County, West Virginia and was placed on the National Register of Historic Places.

Anyway, I never knew any of them. By the time I was six I had only one grandparent, my mother's mother Grandmother (Jenny) Brown. Her father, Edward Day, was born Liverpool, England in 1843. He came to Ohio at a young age and was orphaned by the age of 8. At the beginning of the Civil War, he enlisted as a private for Company F of the Ohio Infantry and fought on the Union side.

Edward married Rachel Carey in 1868 and they moved to Iowa to farm. (Rachel's great-grandmother, Rachel Doane Carey, is a direct descendant of Deacon John Doane, an assistant governor of Plymouth Colony.) Edward and Rachel had seven children, and Grandma Jenny was their fourth.

(From left) Flossie, Lois, Bob, Grandmother Brown (seated),
Xenia, and Louise

Grandmother Jenny lived in Villisca, Iowa, and we visited her often. We went to Villisca for most of the groceries and other supplies we needed and would always stop and see her on these trips. In the summer I stayed a week visiting her and walked to the swimming pool every day for a swim. She was active outside of her house and always had a big vegetable garden and a number of grapevines that she shared with us. I can remember spending time with my Aunt Lois and Aunt Louise on Saturday nights when we were in town.

Grandmothers' house was a small bungalow with a small front porch. She had a porch swing and spent many hours sitting and swinging, holding Effie and me on the swing. She was an excellent housekeeper and for family dinners she always brought candied apples. When the first television sets were available, her children purchased a small black and white unit for her. She became an avid baseball fan. She loved the Brooklyn Dodgers and was upset when the team moved to Los Angeles. In the beginning of TV programming, baseball was on every afternoon. If you were to visit her, you couldn't talk while the game was on.

She was the mother of six children; five daughters and one son. My mother was the oldest child and Louise was the youngest. Louise was in high school when I was in grade school. Robert lived in the area until the early 1940s when he moved to Texas and then California. He finally settled in Oregon. I can remember playing with his three children, Robert, Maxine, and Colleen Brown, before they moved away.

Jessie also lived in the area, and she had three sons. One of the boys, Blake, died when he was nine from an appendicitis infection. The other boys, James and Dean Gourley, were my age, and I often visited and stayed overnight. Xenia, Lois, and Louise moved away from the area. Louise and her family came back from Cedar Falls to visit nearly every Memorial Day.

Grandmother Jenny passed away on July 14, 1969 at the age of 90—one week before the moon landing.

2

MOVING

MARCH 4, 1932

*Picture of the farmstead taken west of the house looking east
up the mud road*

My earliest memory is moving from the house where I was born to the Brown
farm where I lived until I left home to be on my own. It was called the Brown
farm because it was owned by my maternal grandmother, Genevieve Day Brown
(Grandmother Jenny).

Grandfather and Grandmother Brown had purchased the farm in 1902. In December of 1929, Ezra Brown was killed by the enraged Jersey bull while moving cattle to a corn stalk field. Grandmothers' cousin, Jake Carey, came from Nebraska to help on the farm until spring when an auction was held and everything was sold. She then rented the land out and lived in the house until my dad rented the farm from her starting March 1, 1932, and Grandmother moved to Villisca.

Most Iowa farmers rented the land they farmed. Dad rented the farm until 1945 when he purchased the land for $100 per acre. The farm consisted of 160 acres in two 80-acre parcels. This was larger than the average size of farms in the area and was an ideal size for a family operation. One 80-acre parcel was on the north side of the road, and the long side ran north one-half mile with the short side going west one-quarter mile. The other 80-acre parcel was on the south side of the road. This eighty acres was reversed with the short side running south and the long side running west. The house and farm buildings were located in the southeast corner of the north 80.

March 1 was the traditional moving day for farmers because spring was close and planting of small grain usually started in March. Also most of the winter feed supply had been fed to the animals and there was less to move. It took a lot of planning and coordination for those moving. One house needed to be empty or at least things moved to a room so the new farm family could move in. The friends and neighbors of the moving families all helped load things on wagons to take to the new farm. The last things loaded were the beds and heating stoves. When arriving at the new place these items were unloaded first and set up so the house was heated and there was a place to sleep the first night.

Some families moved 20 miles or more and it was a long day traveling with a team of horses. Livestock, such as beef, dairy cattle, sheep, and horses not pulling wagons, were driven on the roads. Someone needed to look out for other cattle being moved and have a plan to keep the herds separated. Hogs and chickens were hauled on wagons or by truck. Some items that were not needed soon, such as farm machinery, excess hay, and corn, were moved at a later date if everyone involved agreed.

Sometimes, if a farm house was empty and no one was moving in that day, there was a big party. At least they seemed like big parties to me, but there were probably less than 30 people. The local musicians played and people danced or just sat on the floor (some brought pillows to sit on) and listened to the music. People brought food and the party was sort of a goodbye to the ones moving and a thank you to those who had helped with the moving. I always thought the empty house was sort of neat. Some of the landowners put a stop to the parties as they thought there was too much damage to their houses.

By the 1940s there were fewer farmers moving to new farms and many more moving off the farm as the purchase of tractors allowed the farm size to increase and fewer farmers were needed to farm the land.

***

The farmstead in 1950

It was moving time for the Narigons. Effie Lee and I were taken to Grand-mother Jenny's house in Villisca to stay while the moving process was going on. I expect Dad took us over by car. Dad had bought his first car in 1925. It was a new Ford Roadster that cost $495. By the time I was born he had a Model A Ford.

After the moving was completed, Grandmother, Effie Lee, and I rode the train from Villisca to Nodaway. (This is my earliest memory). It must have been muddy as Dad met us at the train station with the team of horses and wagon. Or maybe he had planned to purchase some things in Nodaway and needed the wagon to haul them home. Anyway I can remember sitting with Dad on the wagon seat as we headed to our new home.

The house we moved into was modern for the times. It sat on top of a short hill, up a lane from the road. Built in 1915, it was a large square house with a large porch going halfway around the house. There was a full basement with four rooms. The kitchen, dining room (we used it as the family room), a parlor, and a back bedroom were on the first floor. Four bedrooms and a bath were on the second floor.

A large wood or coal-burning furnace was in the basement with heat registers on the first floor and one in the bathroom on the second floor. Dad cut wood in the fall or winter, and we always had a big supply. He also hauled coal from the mines at Carbon or Dickeyville with a team and wagon. The coal was stored in a room in the basement. The woodpile was out in the yard and wood was carried down to the furnace.

There was no electricity so the furnace worked on the theory that hot air rises and cold air sinks. Hot air came out the hot air register and the cold went down the cold air return. Once the fire got burning good the first floor warmed up. The main problem was keeping the fire going at night when everyone was upstairs sleeping.

Dad tried to "bank the fire" at night. To bank a fire, the dampers were closed to restrict the air flow to the fire, and a large amount of wood or coal was added so the fire glowed all night. In the morning when Dad opened the damper and added fuel the fire started very fast, and warm air moved up into the first floor. If it was a really cold night, it might freeze in the upstairs rooms. In the winter the best place to dress for school was over the hot air register in the family room. Effie and I fought over the register.

A carbide gas lighting system was installed when the house was built. This system no longer worked or quit working shortly after we moved in, so we used kerosene oil lamps for lighting. A telephone was on the wall in the kitchen. Since we had no electricity, it was powered with dry cell batteries. The lines were owned by the users, and it was their responsibility to keep the lines working. After every storm the farmers would get together and walk the lines clearing any trees that had fallen on the wires.

Everyone was on a party line with different rings for each person. I think our ring was two shorts and one long. The switch board operator was in Nodaway, and it was her job to answer your ring and place your call. If someone died or there was some other emergency, there would be a long ring and everyone on the line would listen. You could talk to people on your party line by ringing their ring. The operator's family lived in the house with the switchboard so someone was on duty 24-hours-a-day.

The house did have a bathroom on the second floor with a stool, tub, and sink. The only problem was getting water pressure to the second floor. In the basement there was a hand piston pump that was pumped back and forth to bring water from a cistern to a large tank. This tank was sealed, and as the water was pumped into the tank the pressure increased. If you pumped long enough, the pressure got up to eighty pounds.

If you ran upstairs there was enough pressure to flush the stool and still have water to take a bath. The water was heated by passing through a small tank in the kitchen cook stove. If there was no fire there was no hot water. The cistern, a round hole like a well, was filled as rainwater ran off the house. The water was not clean enough to drink but was used for all other purposes.

The drains that carried the waste water ran out to the north side of the house and halfway down the hill to a cesspool. This was a big hole in the ground that allowed the water to purify some before seeping into the ground. This system was outlawed in later years when all houses were electrified and cold pump water and raw sewage became a problem. When the house was modernized with running water a septic system was installed. We also had an outhouse that was located out in the north part of the yard. It got lots of use.

It was a great house and at the time one of the best in the neighborhood. My folks lived in the house for many years and my brother Don still owns the place. (Editor's note: as of 2025 the house is owned by Don's son, David O. Narigon.)

There were several other buildings needed for the farming activities. These included a chicken laying house, a chick brooding house, a corn crib and grain storage building, a hog house, and a nice barn. The barn has stanchions for 12 milk cows, four double-horse stalls and a single-horse stall, and an open livestock pen along one side. The horse stalls and dairy area had a flat ceiling, and hay or straw was stored on the top. The center of the barn was used for hay storage.

3

SCHOOL

SEPTEMBER 1934

The wooden bus we rode to school in 1934. Floyd Agnew was the driver. Photo reproduced from Nodaway Iowa – Past and Present (1976)

I got out of bed early this Monday morning. I dressed in my new shirt, overalls, and shoes. I had gone barefoot all summer and the shoes didn't feel just right, but my mother said they would loosen up in a few days after I had worn them.

After a quick breakfast, Effie Lee and I walked down the lane and waited. Finally a funny-looking green vehicle turned into our drive and the door opened. We got on, and I was on my way to school.

The bus was made of wood and painted green. It had a bench along each side to sit on. The windows were wooden-framed and rattled as the bus bumped along on the dirt roads. The bus was crowded when everyone got on. I can remember how the O'Riley twin girls would fight over holding me on their laps. They were five or six grades ahead of me. I was only four years and five-months-old and small for my age when I started school.

Nodaway Consolidated School (2012)

The bus stopped and let us off at the Nodaway Consolidated School in Nodaway, Iowa. I was fortunate to be able to go to a modern school building. Nearly all the children in Iowa that started to school in 1934 went to a small one-room school house. There was a school house located every four miles so no one walked more than two miles to school. Each school house had one teacher and provided for grades one through eight. Students going on to high school attended a school at the closest town.

The first school in Nodaway was opened in 1869. It was a township school with classes one through eight. A four-year high school was organized in 1905. The first senior class of two students graduated in 1909.

In 1919, five rural school districts voted to consolidate with the Nodaway School District. This made a district of 24 sections of land. (A section of land is one square mile.) In 1922, a new school building was built at a cost of $80,000. This was a modern building at the time. It was three stories with a full basement. In the basement was the gym, locker rooms, a furnace room, and supply rooms. A big stage was along one end of the gym. The gym was used for basketball games, proms, meetings, plays, band concerts, and for recess in bad weather.

The last class graduated in 1959. In July of 1959 the Nodaway School District merged with the Villisca School District. Villisca, as I mentioned, was the county seat and was a bigger town than Nodaway, which was located about eight miles east of Villisca. I graduated in 1946 in a class of nine—five boys and four girls. In my first grade there were 18 students. The next year, in the fall of 1935, the class was divided and some students remained in first grade and the new students started the first kindergarten.

Getting to school was often a real problem. When it rained or in the spring when the frost went out of the ground, the dirt roads turned to mud—deep, deep mud. If the school bus got stuck, some of the farmers along the bus route pulled it out with horses or a tractor. In the winter of 1934 or 1936 there was no school for 30 days because the roads were drifted full of snow. At that time there was no way to move the snow. The only road grading equipment was a horse-drawn blade that did not work well for moving snow.

A train of tractors pull the school bus out of trouble. Photo reproduced from Nodaway Iowa – Past and Present (1976)

In 1934 the road north from Nodaway to Highway 34 was paved. This road was three-quarters of a mile from our house through a field, but it was still more than two miles on a dirt road to drive to the pavement. When Effie Lee and I were in high school we often walked through the field to the Johnsons' and rode with them to school activities. If it was late and dark we stayed the night at their house.

When school started in the fall of 1935 we had new buses that were round and yellow with seats facing the front. There was an entry door in the front and an emergency door in the back.

New buses on parade. Photo reproduced from Nodaway Iowa
– Past and Present (1976)

The school building was divided with grades kindergarten through six on the first and second floors and seventh through twelfth grade on the top floor. The first and second floors were divided into small rooms that housed each grade. The top floor had a large assembly room where all the upper grade students had their desks. In the first row next to the door was the seventh grade. Each year you moved over one row until in the twelfth grade you sat by the windows on the east side of the building. There were small rooms around the outside of the hall where the classes were held. Every 50 minutes the bell rang and you moved to your next class. If you didn't have a class, you remained at your desk and studied. Or you were expected to be studying.

This photo was taken on my 16th birthday with the boys in my grade. Don is on the far left; I am second from right

The enrollment was too small for football but there was girls' and boys' basketball during the winter and boys' baseball in the fall and spring. Basketball was the favorite sport for the students, their parents, and the community. There were usually large crowds for home games, especially if there was a winning team.

Nodaway belonged to a conference of like-sized schools, and we traveled by school bus to the away games. The boys' and girls' teams played the same night with the girls' game always first. The girls' game consisted of six players—three guards (defense) and three forwards (offense). The game was played three-on-three on each end of the court. Players were restricted to one end and could not cross the center line.

When a team scored, the ball was given to a forward of the opposing team at the center line and she started the offensive action in their court. The girls dribbled only once, and there were no three-point shots. The guards could not touch the ball in the forward's hands if they were in the lane under the basket. The game was an offensive battle. A real tall girl on a team could score a lot of points.

Girls' basketball was a popular sport in rural Iowa, and if a team made it to the state tournament the town closed and everyone went to Des Moines and watched the team play.

The boys played the regular five-on-five but without the three-point shots. Some of the gyms were tiny; the circles around the free throw lines nearly met in the center of the gym. The walls were often the out-of-bounds line. One gym we visited had a furnace in the corner with the out-of-bounds line drawn around it. I really enjoyed playing basketball, and the year I was a junior we had a good team. I was a reserve and played guard.

Other activities at school were marching band, concert band, music, and drama. I played the snare drum in marching and concert band. We played mostly marching music and made lots of noise. The marching band traveled to parades, contests, and town festivals. The band won first place at a contest in Pickering, Missouri, and at several other places.

The marching band

When the bus arrived home at the end of the school day we ran up the lane, turned on the radio, and listened to our favorite soaps. Soap companies sponsored the shows so they were called soap operas. Women listened in the afternoon to romance soaps and right after school the kid programs were on. My favorites were "Jack Armstrong, the All-American Boy," "Captain Midnight," and "Just Plain Bill." I could only listen for an hour or Dad yelled at me to get the chores done.

School days were good. There were few problems and I was in trouble only once. One nice spring day when I was a freshman, my buddies and I decided to take the afternoon off. We walked downtown and just hung out for a while. We took a hike out of town to the river and explored the old grain elevator on the way back. We got back to school just in time to get on the bus.

Much to our surprise the school superintendent was waiting by the bus for us. He took us up to his office and called our parents. He informed them that we had skipped school and we needed a ride home. Dad said I could walk home, which I did. He then scolded me about skipping school. A neighbor thought I should be given a hard time about skipping school and wanted to know why I had done such a bad thing. I told him that we were ahead of the teachers with our work and wanted to give them some time to catch up. He thought this was so funny he never forgot it. For years after that when he saw me he wanted to know who was ahead—me or the teachers.

When I graduated in 1946 there were nine students in my class. I finished a distant third in the class. When son Ed graduated from Fort Dodge Senior High there were more than 500 students in the senior class. I chided him for not ranking in the top ten of his class and pointed out that I had finished third. He soon informed me that I wasn't in the top 25 percent of my class as he was in his. I didn't mention my class rating to any of the rest of my children.

4

SATURDAY NIGHT

The biggest social event of the week in the summer was Saturday night in town. Every farmer in the area did their chores early and then loaded up the family and headed for the closest town. We always went to Villisca. The plan was to get there early and find a place to park on the main street. If we were successful in finding a parking place our evening was complete. Now we were able to just sit in the car and watch the rest of the people walk past. There was lots of visiting with friends.

By six o'clock, all the parking spots on the main street were full and the latecomers parked on the side streets and walked up and down the streets. The stores lined the street and as shoppers walked from store to store they stopped to visit. Of course we had the shopping to do but it seemed to me the main objective of going to town was to visit.

We usually tried to get to town early enough to stop at Grandmother Brown's house before going on uptown. If we were late, we left town early and stopped on the way home. When I was real little, the kids stayed with Grandmother while Dad and Mother did their shopping.

As I grew older, I was allowed more freedom in town. At first I ran around in front of the car or sat on the window ledge of a close store. Then I was allowed to walk up and down the block. I met the kids from school, and we walked the streets.

The big treat when I was older was to go to the movie with my friends. I think it cost a quarter and there were two movies. The first was a short B movie, usually a Western, followed by the main attraction. The movie projector had two big reels, and the tape wound through from one reel to the next. The tape broke often, and the lights were turned on while it was repaired. Everyone yelled and made lots of noise until the movie started again. After the movie we had enough money to purchase an ice cream cone for five cents.

About once a month Dad and I needed to get a haircut. The barbershop was in a basement under a clothing store. Steps went down from the street into the shop. Saturday night was the barber's busiest time of the week. Sometimes we waited up to two hours for our turn in the chair. The barber also trimmed beards, shaved faces, and had a bath tub in a small room for those needing a bath. It was a fun place to visit as there was always lots of chatter about what was going on in the community. Some Saturday nights the barber worked until after midnight to get all the haircuts finished.

Mother walked down to the grocery store to buy the supplies we needed for the next week. She didn't buy many things but got salt, pepper, flour, and other baking ingredients. The flour she purchased was in cloth sacks that had different prints on the material. She used the empty flour sacks to make curtains, dresses, and other articles of clothing. The problem was getting enough sacks of the same color and design to finish a project.

In the front of the grocery store was a storage area with shelves to hold the purchases until we drove by and picked them up on the way home. The last stop on the way out of town was the ice house. We picked up a 50-pound block of ice, wrapped it in an old quilt, and put it on the front bumper of the car. When we got home, the ice was carried down into the cave, and it was used to freeze ice cream for Sunday dinner.

For several summers when I was in grade school, the few merchants of Nodaway sponsored free outdoor movies to try to get more customers to spend Saturday night there. We still drove to Villisca to see Grandmother and buy things that weren't available in Nodaway and then drove back for the free movies. The Nodaway lumberyard furnished the bridge planks and cement blocks to make the seats, and there was a large billboard painted white that was used as a screen for the movies. As I remember there was usually a problem with the projector or the film would break. This delayed the showing. In mid-June it was light until nine o'clock and it needed to be dark to see the movie. It was a late night if we stayed for the movie.

Saturday night was the only night stores were open and most people only went to town once a week. If you decided you needed something on Wednesday you waited until Saturday to get it. During the winter, even when I was young, shopping was done on Saturday afternoon and usually by Dad. The rest of us stayed at home. No one wanted to deal with little kids on a cold day in town.

In the late 1940s the big Saturday night shopping started to die out. With better roads and better cars farmers started going to town during the week. When electricity covered the farming area and TV arrived, Saturday night was no longer the time to shop. In a few years all the stores were closed on Saturday night.

***

Sunday

Sundays in the summer were special days. Everyone bathed on Saturday night before going to town. Dad got up early to get the chores done in time to go to Sunday school and church.

When I was older, I was expected to help but in the summer there weren't a lot of chores. Milking the cows, separating the milk, and feeding the little calves were done in the morning. The evening chores on Sunday included a repeat of the morning, plus feeding and watering the chickens, gathering the eggs, and feeding the hogs and cattle that were in the feeding lot.

When the morning chores were finished, we drove to Nodaway for church. It was fun to see my school friends as we didn't get together during the week when school was out for the summer.

When the church service was over, we drove home for a big dinner. If there was any entertaining to do, it was done at Sunday noon and the guests came for dinner.

We often had relatives visit for Sunday dinner. The women worked in the kitchen preparing the dinner while the men sat around and talked. The dinner included fried chicken, mashed potatoes and gravy, or new potatoes and fresh peas, homemade bread, jellies and jams, and vegetables. For dessert there was always homemade pie and homemade ice cream. The men all ate first. The children sat at a special table and ate next. After doing the serving, the women ate. The food was made with real butter, real cream, and fried in animal fat. It was wonderful.

The ice used to freeze the ice cream was purchased in town on Saturday night. The ice was placed in a gunny sack and cracked into small pieces using a hammer or a small mallet. We had a gallon, hand-cranked ice cream freezer. The ice cream mixture was placed in a metal container with a dasher that scraped the side and stirred the mixture as it froze. Ice and salt was dropped in the space between the metal container and the wooden bucket. As the crank was turned, the metal container turned one way while the dasher turned the opposite way. It took nearly 30 minutes for the mixture to freeze and was it ever good! If you took too big a bite, a severe headache would occur. I could never figure out why this happened, but I soon learned to take small bites.

After dinner the men all took a nap on the floor or in a chair while the women cleaned up the dishes and kitchen. Every dish, fork, spoon, pot, and pan had to be washed by hand, dried, and put away. This took at least an hour or more. Sunday was not an easy day for women.

While the men were sleeping and the women working, we kids played outside. We played hide and seek or swung on the hay rope in the empty hay mow, or played ball if there were enough kids and anyone had a ball and bat.

Sometime in the early 1940s Dad purchased a refrigerator. We still didn't have electricity so it was powered by a small kerosene burner. The refrigerator was placed on the screened porch just off the kitchen so the smell of kerosene was not a problem in the house. I'm not sure exactly how it worked but the burner gave off enough heat to run a small motor to circulate the Freon through the coils that cooled the box. Every night we tried a different ice cream recipe that Mother froze in the ice cube trays. She had a lot of recipes for the frozen desserts but none were as good as the ice cream made in the hand-turned freezer.

On the Sundays that we didn't entertain we still had a big dinner. After dinner was over and the dishes were washed, dried, and put away we went calling. It was traditional to go uninvited to visit with a neighbor. The whole family piled into the car and we started out. If there was no one home at the first place we drove on to another place. The visiting was over by four o'clock so everyone could do chores. When we got home Dad always checked the driveway to see if there were any new tracks of someone that had been to visit. Occasionally we met someone in the road who was on their way to visit us.

In the winter the roads were usually bad and it was cold and there were lots of chores to do, so visiting was limited. The only company that came was invited relatives. If there was snow on the ground we had a great time sledding down our lane. If there was no snow, we played card games in the house.

5

CHURCH

Nodaway Methodist Church

We belonged to the United Methodist Church in Nodaway, and I have early memories of going there for Sunday school and the church sermon. Dad and Mother were both Christian and believed in God but were not zealots about religion.

We all bathed on Saturday night to be ready for church on Sunday morning. When I was little, we had baths in a big tub in the kitchen; starting with the littlest kid and working up to the oldest. In the summer Dad filled a barrel with water that sat in the sun in the yard. By evening the water was warm, and we showered out in the yard using a gallon bucket with holes in the bottom. We were behind the house so no one could see us if they drove in the drive.

We only did the necessary chores on Sunday, but Dad wasn't opposed to doing field work if he was falling behind in his work. We went to church nearly every Sunday except in the real cold weather, or when the roads were blocked with snow, or if it was too muddy to get out with the car.

Church did provide a social activity for the family as well as some religious training. The Sunday school classes were small and nearly all the students were the same boys and girls that were in grade school with me. The high school students belonged to a youth group that met on Sunday evenings for various kinds of activities including Bible study.

Sunday school was held the first thing on Sunday morning followed by the worship service. The minister had more than one church to serve so sometimes there was a delay waiting for the minister to arrive. One of the church members was in charge and we sang songs until he arrived.

The church was started in 1864 by Marshall Bullock. The first church services were held in the White school that was located just south of the cemetery. The services were moved to Nodaway in 1869. A new church at the present location was built in 1903. In 1922 the building burned and the present church building was built at the same location and was dedicated in October of 1924. This information was obtained from the booklet *Nodaway Iowa – Past and Present* that was published in 1976.

The church was a modern building for the period. Nodaway did have electricity so there were lights and running water. The heat was provided by a large furnace that burned coal. The janitor arrived early on Sunday morning to get the furnace started and the building warmed up before the people arrived. The sanctuary was nice with large, stained-glass windows on both sides, a stage across the front, and church pews with capacity for more than one hundred people. Most of the Sunday school classes, church dinners, and other social activities were held in the basement.

I can remember as a kid all the different Christmas programs that were held. Each class had some part in the program, and we stood on the stage and read a short verse. As we got older, our parts got bigger. It was always a fun activity after my part was read. Santa arrived and gave us an orange or some candy.

6

HORSES

Dad with the colts

Horses and humans provided nearly all the power used in farming until the late 1930s or early 1940s. A few farmers owned a small gas engine that was used to pump water or run small machines like a feed grinder or corn sheller. A farmer that owned a grain separator (threshing machine) had a big steam engine for the power source. All machinery needed to run a farm was built to be pulled by a horse or mule.

We not only used horses for farm work; but we also raised them to sell. Dad liked horses and every spring there were from one to five new colts born. Dad preferred the Percheron breed, a large docile breed that was developed in France. The colts were born black but changed to gray or dapple as they aged.

One of the largest Percheron breeders in Iowa lived near Corning. He imported stallions and mares from France and always had a big display at the Adams County Fair and at the Iowa State Fair.

Most farmers did not own enough mares to afford their own stallion so they organized a horse group to buy a good stallion to share. One member of the group housed the stallion and was in charge of the breeding program. In the spring when a mare was in heat, Dad called the stallion keeper and he rode a horse and led the stallion over to service the mare.

The little colts were frisky. You needed to be careful when you were around them as some liked to kick with their hind feet. One day Dad walked into the stall and got a colt up. The colt kicked and hit Dad right in the mouth. It split his lip so badly he had to drive to Corning and have the doctor sew up the lip.

Pedro and me on a nice winter day

Most farmers took great pride in the quality of their horses. One of the greatest horses Dad ever owned was named Pedro. I don't know how he got that name, but I can still hear Dad yell "Pedro" when he wanted some extra effort on a pull. Pedro was so great because he was smart and easy to work with. He was used on the hay rope to pull the hay up into the barn. When the man in the haymow yelled to drop the hay, Pedro turned back to the barn. When Dad trained the two-year-old colts to work, they were always hitched next to Pedro. If they started acting up or didn't want to pull, Pedro bit them. He was so big and strong he held the colts in place if they wanted to run.

In the winter there wasn't a lot of work for the horses to do. Every day Dad harnessed a team to haul and spread manure, haul a load of hay or corn, or go to the coal mine for a load of coal, but it wasn't heavy work. Before the real heavy spring work started, the horses needed conditioning to prepare them to work all day. The hardest work was spring plowing. Dad changed teams at noon until the horses were in good working condition.

In 1939 Dad got his first tractor, an International Regular. He traded some horses as part of the cost of the tractor. The tractor had steel wheels. The rear wheels were the drive wheels and had sharp lugs that stuck in the ground to give the tractor traction. The front wheels were close together and were used to steer the tractor. If you hit a rough spot the front wheels turned quickly and spun the steering wheel—cracking your fingers if you weren't holding on tight.

The International tractors had a four-piston, four-cycle engine. The John Deere had only two pistons and a two-cycle engine. The John Deere engine had a distinctive "pop, pop, and pop" sound when running. There was a big debate among farmers as to which tractor company made the best tractor.

The tractor had a pulley on the side that was used to power small equipment like water pumps, feed grinders, and grain elevators. A belt ran from the pulley on the tractor to a pulley on the machine. The pulley ran while the tractor was stationary. Tractors were later equipped with a rear power shaft, and this expanded the use of the tractor.

Soon all the farmers owned tractors and by 1950 there were few draft horses left on Iowa farms. One farmer said, "I sold my last team when I discovered the only thing I was using them for was to clean out the horse barn."

Like all farm boys I soon learned to drive the tractor. At first I just pulled things around the farm lots, but by the time I was 12 I drove all day in the field. My job was to pull a horse-drawn machine with a tongue shortened to fit a tractor. Dad would ride on and operate the implement. The tractor and I had replaced the team of horses. When crossing a ditch or gully I slowed down the tractor but often sped up before the machine crossed; this gave Dad a jolt and he yelled at me. I learned real soon to slow down until he crossed.

Freddie and me in front of the house.

Freddie

In 1939 or 1940 Effie Lee and I got our own pony. His name was Freddie. He was a little tan and white Shetland gelding.

He was about four feet tall and real round. He was so fat that we tried to keep him on a diet but he never got any smaller. We didn't have a saddle so we learned to ride bareback. In the summer one of us rode up to the mailbox each day to get the mail. It was a quarter of a mile to the box so it made a nice short ride. Freddie liked to see "ghosts" in the grass. As I rode along, he jumped sideways like there was something going to get him. If you weren't alert, off you went, and if you dropped the reins, he trotted home without you.

We learned to ride him in the house yard. It was fenced in so if you fell off he couldn't run away. Freddie soon learned that if he went under the clothes line, he could scrape off anyone riding. He was a smart and ornery little horse. He really didn't like to work, but he was fun to have around.

Freddie's life was short and it was a sad day for all of us when we found him in the pasture with a broken leg. A big horse must have kicked him. Our veterinarian came out, but there was nothing he could do to save Freddie's life.

I'm on Dick and Effie Lee is riding Daisy.

Dad purchased two riding horses named Dick and Daisy. They were pretty, brown and white spotted horses. I think Don and Ellie rode them more than Effie Lee or I did. Daisy was bred to a pretty Palomino stallion but her colt was stillborn. That was the last colt born on the farm.

Don riding Daisy in front of the new barn in 1945

7

DAIRY CATTLE

Me with one of the calves I fed

It's 5:30 p.m. on January 18, 1939, and I am in the barn sitting on a one-legged stool milking a cow. I really didn't like milking cows, but I always watched Dad milk and I thought it looked like fun, so as soon as I was strong enough to squeeze the teats hard enough to get milk I had a job. Milking in the winter was a warm chore. The cows gave off a lot of heat, and your hands were warm while milking.

When milking in the summer, it was hot and the flies were always biting the cows. The cows swatted at them with their tails, hitting everything within reach, including your face. Nothing like a swat in the face with a cow's wet tail to get your attention. All the farm cats gathered around behind the cows at milking time because they knew this was their feeding time. Small dishes were placed around to fill with milk for the cats, but it was more fun to squirt the milk right at the cat's mouth. They soon learned to sit there and drink the milk as fast as you could squirt it.

Dad held the milk bucket between his legs while he milked, but I had to place mine on the floor under the cow. When the cow kicked or moved I had to balance on my stool, grab the bucket, and get it out of the way of her feet. The worst was when the bucket was almost full and the cow kicked it over. You could really understand the saying "don't cry over spilled milk."

The milk stool was quite unique and simple to make. A 2x4 board was cut to the length needed to fit the person milking—between 10 and 16 inches long. The 2x4 was stood on end and a 1x4 about 10 inches in length was nailed in the middle to one end and you had a stool. It looked like a capital "T."

When you finished milking a cow, you carried the milk to the separator room. In our old barn the separator was in a little building approximately 50 feet from the barn. When the new barn was built, a room was added at the end of the dairy area for the separator.

The cream was the only part of the milk that was sold. The separator was used to separate the cream from the milk. The separator was about four-feet-tall and had a large bucket-like container at the top. The milk was poured through a milk strainer to catch all the hay leaves, flies, and other small items that might get in the bucket while you were milking. The container had a valve you turned to allow the milk to flow through the separator. A float held the milk back so it wouldn't overflow.

To separate the cream, I turned the handle of the separator at the proper speed. There was a little bell on the handle, and when the bell quit ringing it was time to turn on the milk. The milk flowed down through a cylinder that had a number of thin disks.

Since the cream was lighter than milk it spun up and flowed out of the top spout into the cream bucket. The skimmed milk ran out the bottom spout into another bucket. Skimmed milk was fed to the hogs. Whole milk was saved for serving in the house. We always had a supply of fresh milk to use.

Washing the separator was a big job. After the evening milking the separator was rinsed with water and used again the next morning. Every day after the morning milking the separator was taken apart and carried to the house for washing. There was a little hanger to slide into the disks to keep them in order so they always fit the spindle.

The dairy cows were housed on one side of the barn. I think we had 10 or 12 stanchions that each held a cow. A cow stuck her head through the stanchion to get feed, and I closed it around her neck. She was then unable to back out as the stanchion held her head until we were through milking. The cows learned where they were to stand and were always in the same stanchion.

Just behind the cows was a gutter in the floor a foot wide and six-inches-deep. This caught the manure, and we used a scoop to clean out the gutter. In the winter the cows were kept in the barn all night and most of the day. The rest of the year the cows were let into the barn for feeding and milking and then turned out to pasture.

One of my early jobs was to go out into the pasture and drive the cows back to the barn. I was always barefoot in the summer so I needed to be careful where I stepped as there was lots of cow manure around. I also needed to be sure a cow didn't step on my foot. I had a lot of sore toes in the summer from being stepped on.

Nearly every farm had at least one dairy cow to provide milk for the family, and most of the farms had several dairy cows. Selling cream was one source of cash for the farm family. The cream and egg money was used to buy weekly food supplies, clothing, and other necessities.

Many farmers had dairy breed animals: Holstein, Jersey, or Guernseys. These breeds were developed to produce milk and were short of muscle. Dad had Milking Shorthorns, a dual-purpose breed. While the breed was not the greatest milk producer, the calves were meaty and sold for a lot more money than the dairy breed calves. Dad purchased one or two heifers every year from a good

Milking Shorthorn breeder to add to the herd and then used a beef Shorthorn bull to sire the calves.

A cow's gestation period is nine months. Most of the cows were bred to calve in late May and June so we had fewer cows to milk during the corn planting season. When a cow gave birth, she would freshen (produce milk). The cow and calf stayed together for two or three days so the calf nursed the first milk produced (colostrum) after parturition. This milk was high in protein and antibodies that fortified the calf's immune system.

After three days the calf was weaned from the cow. The cow was milked twice a day for the next nine months and then dried up so she stopped producing milk. The calf was moved to a calf pen with the other calves and fed milk twice a day from a bucket. To train the calves to drink out of a bucket I let the calf suck my fingers, then pushed the calf's nose down into the milk, and in a few days the calf learned to put his head right in the bucket and drink. This was light work, so I was young when it became my job to feed the calves while Dad did the separating.

To feed the calves, I carried a bucket of milk to the calf pen and poured about a gallon of milk into the calf pail for each calf. The calves stuck their heads through holes in the fence. I placed the bucket in front of a calf and let them drink. I had to be alert to make sure I didn't feed the same calf twice. The bigger calves always tried to push the little ones back.

Having dairy cows really kept the farmer at home as the cows had to be milked twice a day. We milked the cows the last thing at night and again the first thing in the morning. If you went someplace in the daytime, you had to be home in time to milk. If you were going some place in the morning, you had to milk first. People used to want to farm to be their own boss, but I always felt the milk cows were the boss.

Today there are no longer small dairy farms. Most of the milk is produced in large mechanized facilities owned by large corporations. Some have more than ten thousand cows and milk twenty-four hours a day.

8

BEEF CATTLE

One of our herd bulls

Beef cattle required the least amount of care of all the livestock we raised on the farm. When I was little, Dad had 15 or 20 cows and a bull. The bull was also used for the dairy cows. At that time most of the beef cattle were Hereford, (red animals with a white face), Angus, (all black in color), or Shorthorn (they could be all white, all red, or a roan color). We had the Shorthorn breed which worked well with the Milking Shorthorn dairy animals. These were two distinct pure breeds of cattle.

The beef cattle were in the pastures all summer and the only care they really needed was to make sure they stayed in the pasture and to pump water for them when the ditch dried up. The cows had their calves in the spring, and if the weather was bad we kept the cow and the calf in the barn for a few days. A local veterinarian came out to the farm to vaccinate the calves for several diseases when they were small and the males were castrated at the same time.

The cows and their calves spent the summer grazing in the different pastures. If there was little rain, the pastures became short of grass and Dad turned the cows out on the road where they ate the grass along the road ditches. I was the guard on one end of the road and Effie Lee watched the other end to keep the cattle at home. In the fall the calves were weaned from their mothers and shut in a lot to feed. The cows spent the winter in the corn stalk fields and were fed hay.

In 1939 or 1940 Dad went to a registered Shorthorn sale and purchased some bull calves for my 4-H project. One of the calves was a small compact animal and we decided to keep him for our herd bull. Registered cattle were ones whose pedigree had been recorded with a breed association and the paper listed the names of all the parents going back several generations. Dad then bought some registered Shorthorn heifers and we were started in the purebred business.

Dad sold some bulls for breeding and steers for 4-H show steers. Elinor exhibited the Champion Shorthorn steer at the Aksarben Livestock Junior show in Omaha with a steer that Dad raised. Aksarben was Nebraska spelled backwards and was the name used for the big center and race track in Omaha.

The steers and heifers from the beef and dairy cows not kept to replace old cows were fed corn and hay until they were ready to sell for beef. We butchered one or two every year for our own meat. The beef animals were usually shipped by truck to the livestock yards in Omaha.

Dad drove out to the yards to see the animals sell. They were consigned to a commission firm that owned some pens in the yards. The salesmen working for the firm fed and watered the cattle when they arrived and had them ready to show the packer buyers. The buyers for the packing companies came through the pens and bid for the cattle. If the buyer and salesman could agree on the price, the cattle were driven to a scale and weighed. The salesman took the weigh ticket back to their office in the exchange building and Dad got his check with the commission fee taken out. The meat-packing plant sent their check to the commission firm.

This was a good way for farmers to sell their animals. Dad was a long-time shipper to the Omaha yards and received an award for being a 50-year shipper. At one time 90 percent of all cattle and hogs were sold through stockyards. The big shipping day was on Sunday so the animals were there for Monday morning's sale. Trucks were often lined up for miles waiting to unload the animals at the Omaha stockyard. At one time it was the second largest stockyard in the country with Chicago being the largest.

The large meat-packing companies were located just outside the yards, and the cattle and hogs were driven to the plants by way of an elevated deck. Some days there were more than 10,000 animals sold at the yards. The radio reported how big a run was expected and the market price so farmers could decide when to ship their animals.

I can remember my first trip with Dad to the Omaha stockyards. We did the chores early and were on the way before the sun came up. As we drove along there were no lights in the farmhouses. One by one the lights started turning on. By the time we got to Omaha the sun was up. We made our way on the walkway over the cattle pens to the commission firm's lots and found our cattle.

The commission man said the market was lower but Dad's steers looked good. He thought they would bring the top price. We watched as a buyer came in the pen and looked the steers over. He made a bid but the salesman said he needed more. The buyer left, and the salesman told Dad he would be back as the market was moving higher. Sure enough the buyer soon returned and bought the cattle at the asking price.

Dad was really pleased, and we made our way to the exchange building. The exchange building was about ten stories high and was the location of all the commission firms' offices. Each office had one or two secretaries, a firm manager, and the space for the salesmen. The large firms might have as many as 15 people working for them. As a kid, I was really impressed with the elevators in the tall building. There were three elevators. Each one was operated by a person, usually a woman, and another woman stood outside the elevator and snapped a stick when the elevator was full or if no one was waiting to ride up. The operator then moved the elevator up and stopped at each floor as the floor numbers were called out.

Big feed lots that supplied a large number of fat cattle came into existence in the 1950s and many farmers quit feeding cattle. Today, very few Iowa farmers raise feed cattle. As packing companies modernized, they built their plants away from major cities and started buying the cattle direct from the feeders. This type of marketing rapidly took over and the livestock yards soon went out of business. Houses and other buildings are now located in Omaha where the livestock yards once flourished. This was another drastic change in agricultural practices.

9

RAISING HOGS

Photo of piglets by RoyBuri courtesy of Pixabay

In the 1930s and '40s hogs were the primary income producing livestock on Iowa's farms. They were called the "mortgage lifter" because most of the land in Iowa was paid for with proceeds from selling hogs. The reason they were so profitable is that they were so prolific. If a farmer owned 10 sows and a boar, with good luck he could sell at least 150 fat hogs in a little over a year in two farrowings (two litters per sow).

One year Dad had his picture in the paper when he marketed 98 pigs from nine sows in just one farrowing. His market average was 10.9 pigs per litter. The average litter size marketed was between seven and eight.

Raising hogs however required a lot of labor, facilities (sheds, feeders, and other equipment), and planning. The market was always the highest when there was a low supply of hogs. So the plan was to have hogs ready for market before everyone else had their hogs ready to sell. To do this, one had to farrow pigs earlier in the winter, take better care of the pigs, or have a faster gaining breed.

Dad always had the Hampshire breed. They were black with a white belt around the hog just behind the front legs. The Hampshire wasn't the fastest gaining breed but was known for their mothering ability and large litters.

Gilts (female pigs) would be selected from rest of the fattening hogs in June. Dad always cut a notch or notches in the ear of the gilts from the largest litters. He did this when the pigs were just a few days old and the notches identified the pigs for life—unless they got in a fight and had their ears chewed up. The notches placed in different parts of the ear corresponded to numbers so each litter and each pig could have a different number.

The gilts were moved to a pasture lot where they could get plenty of exercise and a low fat diet. They grew a big frame with little fat—ideal for a brood sow. In September or October, a boar (male hog) was turned out with the gilts for mating. The gestation period for hogs is 112 to 114 days.

It was always a fun trip to go with Dad to purchase a new boar. We usually bought a purebred Hampshire and there was a farmer near Lenox that raised a large number of Hampshire hogs. He had an auction every spring and fall to sell gilts and boars. Dad, a neighbor, and I drove to Lenox for the auction. The auction was held in the sale barn, which is a building that was built just for livestock auctions. There was a small ring in the middle where the boars were displayed.

The buyers sat on amphitheater seats and bid on the animals they wanted to buy. The auctioneer was loud and banged the bench in front of him. He talked so fast you could hardly understand what price he was calling. The boars were brought into the ring in groups of three or four. Someone described the animal's pedigree and told how good they were and the bidding started. The buyer had the option of taking all the boars in the ring or his choice at the bid price.

One night, a farmer sitting next to the ring fell asleep. The auctioneer and the ring men were laughing at him for being asleep. The next group of boars came in the ring and the auctioneer ran up the bid and yelled really loud "sold!" The man woke up and the auctioneer pointed to him as the winning bidder. The man got up, walked out into the ring, and picked out a boar. He said "I have sure purchased a lot worse boar than this one when I was awake." Everyone in the sale barn really laughed.

Our farm had three types of farrowing buildings. One was a central farrowing house. This building had fourteen pens that were approximately six feet-by-six feet. There were seven pens on each side with a four-foot wide alley in the middle. The roof had glass windows placed so that in the afternoon the sun would shine right in the middle of each pen.

The little pigs really liked to lie in the sun. This also helped with vitamin D for the pigs. The sow's body heat helped heat the building, and there was a wood burning stove in a corner to fire up on a cold night if any of the sows were ready to farrow.

Another type of farrowing building was a hexagon-shaped building. The pens were pie-shaped. They were wide on the outside and narrow in the center of the building. A wood stove was placed in the center, and the little pigs soon learned to stay near the stove. There was a small door in each pen to let the sow in and out. A walk-in door and small alley went to the stove. This building was never popular with swine producers. It was hard to get in the pens with the sows, and a lot of buildings burned when a sow broke into the stove area and upset the stove.

The three-pen portable house was the most popular. These houses could be moved with a team of horses and put in a new location each year. This was one way of controlling swine diseases. The buildings were eight-feet-wide and eighteen-feet-long. Each building had three pens that were eight feet-by-six feet. The front of the building was about three-and-a-half feet high. There was a small door in the front to let the sow in and out. The roof sloped up about six feet to the peak and sloped down to the back of the building. There was a door on the front of the roof that opened to let the sun shine in on sunny days. You used this door to climb into the pen to check the pigs and clean it out.

The cold was the hardest thing on newborn pigs. They were born all wet and could chill quickly and die. Someone checked the sows every hour to see if any sows were farrowing to rescue the little pigs. You could usually tell when a sow was getting ready to farrow as she would build a nest with the straw in the pen and start dripping milk from her teats.

Once the sow settled down, she shelled out the little pigs. If it was my turn to watch, I grabbed the little slimy pig, dried it off with a towel and placed it by the mother's teat. If it started nursing right away, it was off to a good start. If it was a cold night we dried the pigs off, placed them in a basket, and took them to the house and put them by the wood stove. This kept them warm and they could go several hours before they needed to nurse.

Baby pigs were born with sharp, long teeth (tusks). While nursing they fought with each other and often cut the side of each other's head with their little tusks. The cuts could become infected. To prevent this, we caught the little pigs and cut off their tusks with pliers. The ears were notched at the same time.

The next danger for the new born pigs was being crushed by the sow when she would lie down. Lots of little pigs died this way if the sows were not good mothers. Keeping the sows for only two litters helped to control the loss of little pigs because the younger sows were smaller. The sows grew as they got older and took more space when they lied down. In the late 1940s, the farrowing crate was invented. The sow was confined in the crate so she was unable to turn and the loss of pigs due to crushing was reduced.

When the pigs were seven or eight weeks old the males were castrated. It was my job to catch the pig, lay it down on its side, and hold it there with my knee on its neck. I held the front and back legs together and Dad did the job. Of course the pig squealed and the old sows were raising a fuss outside the shed. It made for a very noisy day.

At 10 to 12 weeks of age the pigs were weaned from their mothers. The sows that farrowed in the winter were kept to raise another litter in the fall. The fall sows were sold and new gilts were selected to start the process again.

We caught the pigs one more time to vaccinate them for hog cholera and erysipelas (a skin infection). Again my job was to help Dad catch the pigs. I caught the little ones and Dad caught the big ones. We held them with one front leg in each hand with their back toward us. The vet put a shot in each of the pig's armpits. This was really noisy and usually dusty as pigs liked to roll in the mud. The mud stuck to the pigs until they got to the sleeping area. When the mud dried you had dust.

The pigs, now called shoats, were ready to fatten for market. They were kept in a dirt lot and fed corn and a supplement containing vitamins and minerals. Before we got the tractor, a man came with a grinder on a truck every two weeks and ground corn for the hogs.

We often fed a mixture of milk, water, and ground feed called slop. To feed the pigs you dipped the slop out of a barrel and poured it in a trough. If you went into the pen with the pigs, they always tried to knock you down to get the slop. By putting the end of the trough under the fence we poured the slop in the trough without getting into the pen.

This was called "slopping the hogs."

There was a big water tank in the lot fence that allowed livestock to drink from the two lots on either side of the fence. An underground pipe ran from the tank to the well across the road. A windmill pumped the water to the tank. If no one was there when the tank got full it ran over into the hog lot. In the summer the hogs really liked to play in the wet area.

A waterer was attached to the tank in the hog lot. The water flow was controlled by a float, but on a hot day the hogs splashed the water out and there would soon be a big mud hole with the pigs swimming in it. Each year Dad hauled in fresh dirt to fill the hole and the pigs went back to work digging a new hole.

Gilts and sows not in the fattening lot got water out of a hog waterer that was like a barrel with a drink hole on two sides. The hogs stuck their snouts in to drink. Water was hauled to the waterer in barrels on a wagon. The water was hand pumped into the barrels. In the winter an oil lamp was placed under the waterer to keep the water from freezing.

Hogs loved to root in the ground with their snouts and, unless prevented, they often dug up the entire pasture. The only way to stop this digging was to place a ring in the sow's snout. As soon as I was strong enough to squeeze the pliers shut with the ring, I had a job.

Dad caught the sows with a snare around their snout. A snare was made with a piece of rope about eight-feet-long and a ring was tied to the end of the rope. The rope passed through the ring making a loop. You dropped the rope under the sow's chin and into her mouth. The sow pulled back on the snare, squealing as loud as she could. I placed the ring over the tip of her snout and squeezed the pliers, closing the ring in the snout.

If the sow jerked at the right time I missed and Dad yelled. He yelled a lot.

When the sow tried to root in the soil, the ring jabbed the nose and the sow stopped rooting. Ringing sows' noses was another job I wasn't crazy about doing.

It usually took at least six months for the pigs to reach market weight (220 to 240 pounds). Most farmers had their pigs farrow in late February or early March when there was less cold weather to worry about. The big run of spring hogs to market started in August and September and as the supply increased the price dropped. The goal for us was to have the hogs ready in July and get a higher price. We started by sorting off the biggest ones and hired a truck to haul them to the market in Corning.

"The hogs are out!" That was the battle cry, and everyone available was to race out the door to the farm lot to round up the wayward animals. Swine were considered by many as the smartest of all farm animals, but when it came to getting them back into their pen they really played dumb. They could find a tiny hole in the fence to get out but could never find it to crawl back in. When we opened the gate someone had to stand guard to keep the pigs in the pen while the pigs outside the pen were corralled back in.

Dad was yelling, Shep (the farm dog) was barking, and I was always moving the wrong way. Eventually the hogs were captured only to run another day.

Electricity arrived on our farm in the fall of 1946. Heat lamps were then used to keep the pigs warm. Electricity also powered motors to move feed. It wasn't long until heated and ventilated farrowing houses were built. Nursery buildings and confinement finishing houses soon followed. Today most hogs are raised by large corporate farms in total confinement.

The early hogs were raised for their fat as well as the meat. The fat on the pig's back measured up to two-inches-thick. The fat was used for many industrial items. At home it was rendered into lard and used for cooking and making soap. The term for market hogs was fat hogs.

Industrial needs changed and today's hog is produced for meat only. The market term today is lean hogs and the back fat may measure less than a quarter-inch. Hogs raised today in the confinement buildings would not survive in the conditions of the hogs I helped raise as a boy on the farm.

10

CHICKENS

Photo submitted by David O. Narigon

In 1935, nearly every farmer, if not all, raised chickens. Some farms had just a few that ran all over the farmyard and lived off spilled grain, insects, and weed seeds.

Most farms had a small number of hens and a few roosters. They self-multiplied each spring as the hens sat on eggs, hatched, and raised the little chicks. The number of chickens depended on how many escaped the farm cats and other predators. They provided fresh eggs and meat for the farm family.

We raised chickens not only for the eggs and meat they provided, but also to add cash income to the farm by selling eggs. It was an exciting time when the baby chickens arrived in the spring. Dad picked them up at the hatchery in Villisca. If they had been ordered from a different hatchery, the mailman delivered them to the farm.

Baby chicks came in a specially made chick-shipping cardboard box. The box was approximately 24-inches-square and six-inches-deep. The sides were lined with small punch-out holes. The number of holes punched out depended on how hot or cold it was and how much ventilation was needed. The box was divided into four sections with each section holding approximately 25 baby chicks.

When the chick was hatched the yolk (yellow part of the egg) remained in the chick and served as the food supply for several days. This allowed time for the chickens to be shipped by the U. S. Postal Service. If you visited a post office in the spring it was full of boxes of baby chickens.

In preparation, the brooder house had been cleaned and disinfected. The brooder stove was also cleaned and started, and mulch bedding was spread on the floor. We were ready for the chickens' arrival. The brooder stove burned kerosene as the heat source, and it had a large metal hover that hung about 12 inches off the floor. The hover acted like a false ceiling and held the heat down on the chickens.

The chickens arrived. If it was a Saturday or after school I was there helping.

I reached in the box and lifted out a chicken, dipped its beak in the water, made sure it drank, and then turned it loose. (Chickens in those days got a lot of individual care.) I then scattered some chick feed on large pieces of paper. I scratched my finger on the paper. The sound brought the chicks running, and they started to eat the feed. The sound was sort of like the sound the old hen made when she would scratch and call the newly hatched chickens.

We filled small chicken feeders with feed and placed them around the brooder house. The feeders were small metal trays with a lid that slid on. The chickens ate through small round holes in the lid. The water dispenser was a quart jar with a cup-like round trough that screwed on the jar. When the chickens drank, air bubbles went into the jar and allowed more water to fill the trough.

One of my early chores was to fill the feeders and waterers for the chickens. The chicken feed was stored in the brooder house and Dad or Mother would carry a bucket of water for me to use. The first few days, I needed to be very careful where I stepped. If I stepped on a chicken, it was goodbye chicken. Dad or Mother filled the stove with fuel and made sure the temperature was right for the chickens.

When I was big enough, I carried the water to the brooder and the laying chicken houses. The easiest way to carry water was to carry a bucket in each hand so I could keep my balance and I wouldn't spill as much water. In the winter, the water that I splashed on my pant legs froze and really kept my legs cold while I was doing chores.

A girl I met at college said, "You could always tell if a girl was from a farm by the way she walked. She had her hands out like she was carrying two buckets."

We usually got the chickens in the early part of March. We wanted the chickens for egg production so we always had the White Leghorn breed. This breed was noted for its high egg production, but wasn't very meaty and was slow to mature. Our goal was to have fresh fried chicken on Dad's birthday on June 8.

Dad purchased a few chickens from the White Rocks breed just to eat, but we still had to eat a lot of the leghorns. Roughly 50 percent of the leghorns were roosters, and we really only needed the hens.

In later years we purchased only the female chicks as the hatchery hired a person that determined the sex of the chicken when it was hatched. I never understood how a person sexed a chicken by looking up its butt. I guess you had to know what to look for.

The baby chickens grew fast and in about six weeks they were all feathered out and no longer needed the brooder stove. The stove was pulled up to the top of the brooder house and stored until the next spring. Larger feeders and waterers replaced the little ones used for the baby chickens. Roosts were placed along one side of the house and the chickens soon started roosting at night. The chickens were allowed to roam around the farm and old Shep, our farm dog, kept the foxes away. At first the chickens were shut in the brooder house every night, but they soon started roosting all over the farm.

June 1 was time to start checking the size of the biggest roosters to see if there was one big enough to fry for Dad's birthday. We used a four-foot-long stiff wire with a hook on the end to catch a chicken by the leg. The catching hook was given to us by the hatchery, and they had their name on the wooden handle as advertisement.

On June 8, Dad's birthday, we caught two chickens that were big enough to fry. Mom started some water boiling on the stove to use to scald the chickens to remove the feathers. Dad placed the chicken's head on a block of wood and cut its head off with a small hatchet. (Some people wrung the chicken's neck by holding the head and swinging the chicken around until the head snapped off.) The headless chicken flopped around the yard while all the blood pumped out. The boiling water was poured into a small pail, and the chicken was dipped into the hot water. The feathers were then easy to remove.

This was one of my jobs, and it was an all-day job when we were dressing a number of chickens to can meat. A piece of paper lit on fire was used to singe the little pin feathers off the chicken. The chickens were then cut up into pieces and placed in cold water until Mother was ready to fry them.

Boy! Was it ever good! The first fried chicken of the year! The rest of the summer we had fried chicken every Sunday for dinner and at least twice during the week.

In the fall we started to find little eggs all around the farmstead. These were pullet (young hen) eggs, and we knew it was time to sell the old hens. Once they were gone, the laying house was cleaned and we were ready to catch the young layers and move them into the laying house.

This was not a simple task, as by now the chickens were wild and roosting all over the farm; on fences, in the trees, and in buildings. One year when we were delivering May baskets (another story), Dad drove us to a neighbor's house one mile away, and a chicken was still roosting on top of the car when we arrived.

When it was dusk, the chickens would find a place to roost for the night. At dark the brooder house door was closed. The next morning, all the chickens still roosting in there were captured and carried to the laying house.

A different plan was made to capture those chickens roosting outside in trees and bushes. A dark evening was selected because if it was a moonlit night the chickens saw you coming and flew before you could catch them. Everyone in the house was called to duty. If you were too small to catch a chicken you could at least carry them after they were caught. The plan started with everyone selecting the chickens they wanted to try to catch. Dad gave the signal and we all grabbed as many chickens by their legs as we could. A good catch for one person was three or four chickens. We carried them to the hen house and then went back to a different location for more chickens.

This process would go on for several nights as the chickens sitting next to the ones you caught flew and did not return to the roost that night. Some of the real cagey ones spent most of the winter roaming around the farm until they were caught almost by accident.

The laying house was located just west of the house, close enough to carry water to the chickens and the eggs back to the cave. The laying house was divided into two sections. The front section was the area for the feeders and waterers. The hens spent the daylight hours in this area. There was a large open window on the south side of the building that let in the sunshine and fresh air. A canvas drape was rolled down at night and during cold weather to keep the building warmer. There was no source of heat in the building except for the small oil lamps that were placed under the waterers in cold weather to keep the water from freezing. We had from one hundred to two hundred chickens and they gave off a lot of heat.

In the middle of the building were the rows of nests. They were made of wood, and each nest was about a 12-inch cube. There was a rail along the front for the hens to walk on to get in the nests. One of my early jobs after school was to gather the eggs. I carried a bucket of water to the hen house and eggs back to the cave. The egg cases were stored in the cave and each night the eggs were placed in the case. Each case held 30 dozen eggs and weighed approximately 32 pounds. A truck came once or twice a week and picked up the egg cases and the cream cans.

Gathering eggs wasn't a bad job except in the spring when some of the hens got broody. They were ready to hatch new chickens and stayed on the nests all the time. They were in a fighting mood and tried to bite me when I reached in for the eggs. One spring I had a lot of warts on one hand, and when I reached for the eggs a hen bit one of the warts right off my hand. It bled and bled but in a few weeks all the warts were gone and I never had warts again. I was glad she did it because Mom was going to take me to the doctor for the warts, and I hated to go to the doctor.

The back section of the building was where the hens would roost at night. There was a false floor about four feet above the ground and the roosts were placed above it. The roosts were long two-by-four boards set in slots that held them on edge. The manure would drop on this false floor.

Every so often the floor needed to be cleaned. When I was old enough, Dad always arranged the cleaning to be done on a Saturday so I was there to help. I think this was one of my worst jobs on the farm. First I pulled the roosts up and stacked them in the back of the area. With a scoop I started at the front, scooped the manure, and threw it out a door into the manure spreader. Dad brought the spreader with the horses and left it by the door as it took several hours to get the roost area cleaned.

Part of the problem was there wasn't enough space to stand up straight as the building had a low ceiling. All this time I was walking in chicken manure, but at least I had rubber boots to wear. And I survived.

For several years, to make extra money, we sold hatching eggs to the Villisca hatchery. To do this we had to keep several roosters and have the hens blood-tested for various diseases. During the 1940s the poultry industry changed rapidly and by the time I graduated from high school few farmers kept chickens other than a few for their own eggs and meat. Large poultry farms started, and today all the eggs produced are in large confinement buildings. All the small businesses that supported the chicken industry are gone.

11

GARDENING

From left, Reverend Congable, me, and Don ready to cultivate the garden. Freddie is not very happy—notice his ears laid back.

Today people garden as a hobby, but when I was a boy a big vegetable garden was a must to feed the family. One big advantage to living on a farm during the Depression was the opportunity to grow nearly all our food supply. We had a big garden.

I can remember my folks fiercely debating how much space there should be between the rows in the garden. Dad wanted a wide space so he could use the horses and cultivator to control the weeds. Mother wanted a narrow space so she could use the hoe to keep the weeds under control. She knew Dad was too busy to cultivate when needed, and if she didn't use the hoe the weeds took over.

One year Reverend Congable lived with us and helped on the farm. He was the minister for the Nodaway and Gus Methodist churches. He was single and had time to work on the farm for his room and board. He was a great horseman and rode a horse to visit people. During the rainy season when the roads were muddy he often rode to the Gus church.

He liked to go to horse sales, and one night he found a harness that with a little repair could fit Freddie, the Shetland pony. He also found a small V-shaped harrow and decided we could weed the garden with Freddie and the harrow. The problem was, no one asked Freddie what he thought of the idea.

When we got him all hooked up and into the garden, he decided he wasn't a work pony. He wouldn't drive. As soon as the tugs got tight to pull the harrow he stopped. We tried to lead him but he pulled back and stepped on the plants. Mother had a fit because he was stomping on all the rows and breaking off the plants.

She said, "If we had all been hoeing instead of spending the time working on the harness and fooling with Freddie, the hoeing would have been finished!" Freddie was retired from garden work.

Planting the garden was always fun for me. It seemed we always planted on a nice warm early spring day. Dad fall-plowed the garden so the freezing and thawing of the ground during the winter and spring made the soil soft and easy to work. If we started working the soil while it was too wet it became cloddy and very hard to work.

The garden was located to the west of the house on the top of the hill. It sloped to the north and the south so there was good drainage. If I remember correctly, the garden was at least 100-feet-long and 60-feet-wide. The vegetable rows ran north and south the long way. We started with the early crops next to the house and worked west with the later plantings.

The first things planted were four rows of peas, radishes, lettuce, onions, cabbage, and parsnips. A little later the carrots, beets, green beans, and tomatoes were put in. The sweet corn, melons, pumpkins, and potatoes were all grown on the edge of a corn field. They were cultivated with the horses and didn't require a lot of hand care, except when the potato bugs arrived and we had to handpick the bugs off the plants. There were no insecticides available to control insects at this time.

One of the jokes we heard about was an ad in the newspaper for a machine to control potato bugs. It cost one dollar and came with complete instructions on how to use it. When the machine arrived it was two blocks of wood with the instructions, "Place the bug on block A and hit it with block B."

Planning the garden started in midwinter when we received the seed catalogs. The pictures in the catalog were so great and when summer came, the vegetables never looked as good as the pictures, but they did help in deciding what things to plant and the cost of seed. I can remember that the Earl May and Henry Field's catalogs, both coming from Shenandoah, Iowa, were the most popular. Dad and Mother ordered seed from both catalogs, and by the time I was in high school we sometimes drove to Shenandoah to get the seeds and plants. It was about 40 miles away and was an all-day trip.

When planting started, Dad brought a team with one section of a harrow to smooth out the soil. We used a long piece of twine tied to stakes and stretched it the length of the garden to make a straight row. Mom pulled the hoe along the twine and made a row about two-inches-deep for peas and beans. Effie Lee or I followed along dropping the seeds in the row. Cabbage and onion plants were placed in little holes in the ground but still along the twine to keep the rows straight. Later in the spring the green beans, squash, and the melons would be planted. Tomato plants were set out around the middle of May.

Planting potatoes was a full day's job and always done on a Saturday so everyone was available to help. The potato patch was not in the garden but moved from year to year to a different location on the farm. This was supposed to help control the insects and diseases of the potato. It seemed to me that bugs and blight found them anyway but maybe not as many as otherwise.

We started with at least 100 pounds of seed potatoes. The day before planting, we cut the potato into small pieces and each piece had to have an eye. The new plant started growing from this eye. Sometimes the eyes on the potato were so close that a piece had several eyes. When we were done cutting, we had several buckets full of cut potatoes. Dad, with the team and harrow, smoothed down the fall-plowed soil.

We didn't use a twine to mark the rows here. Instead, Dad hooked the team to the walking plow. (This was a one-furrow plow that a person walked behind while the team pulled it through the ground.) He plowed a shallow furrow across the patch and the potato eyes were dropped into the furrow about 12 inches apart. Dad complained if you dropped them and they bounced out of the furrow, or if you didn't keep the the potato eyes in a straight row. He then plowed the furrow shut, making a mound over the potato eyes.

The potato patch was cultivated with the team and a one-row cultivator. The horses walked between the rows and the cultivator shovels cut the weeds and rolled dirt over the weeds between the potato plants. The only hand work was to occasionally hoe out any weeds that grew in between the potato plants.

Sweet corn was planted on the edges of the cornfields. Dad used the horse-drawn corn planter to plant the seed. He planted several times during the spring to extend the sweet corn harvest period. Fresh sweet corn-on-the-ear was a real treat. To eat the best corn, the old saying was, "Have the water boiling on the stove. Walk to the sweet corn field, pick the ears, and run, not walk, back to the boiling water."

The first vegetables out of the garden were radishes, onions, and lettuce. Fresh vegetables really tasted good. The best meals were in June when we had new potatoes and peas and fried chicken. We had lived since the fall freeze eating canned or dried food. We couldn't run to the market and buy fresh like we do today.

The first big crop to harvest was the peas. Dad and Mother were always early risers and they believed as much work as possible should be completed during the coolest part of the summer day. We were informed the night before what was to be done the next day (pick, shell, and can peas).

I think I was about six when I started picking, before that I was probably around and in the way but not doing much. To pick peas, you had to choose the pods that were full, round, and hard. The peas did not all mature at the same time. Some were ready to pick that day and many more were picked in the future. This wasn't just a one-day job but lasted several weeks.

The picking took two or three hours, and we filled several three-gallon buckets of peas ready for shelling. We sat in the shade and shelled the peas. To shell, I opened each pod and rolled the peas out into a pan. The peas were washed and "blanched" (boiled in water), and then placed in glass quart canning jars. Mother used a pressure canner to cook the jars. The pressure canner held the steam in to increase the temperature and sterilized the peas quicker than using an open canner.

The early jars I remember had zinc lids and a rubber gasket that went around the jar and sealed when the lid was tightened down. Later the bands and flat lids were invented. One just purchased new lids each year and reused the bands that screwed on the jars.

Green beans

This same process was used to harvest green beans. They were snapped and not shelled. Tomatoes, beets, and corn were all canned in glass jars. I think we probably had around 100 jars of each along with jars of soup, meat, and mixed vegetables. The jars were all stored on shelves in the cave, sometimes called a root cellar.

Harvesting the cabbage was another big job. We carried in the cabbage heads, washed them, cut out the core, and sliced them into a big clay jar. A cabbage cutter was used to slice the cabbage. It set across the jar, and as the cabbage head was pushed across the knife the slices fell into the jar. Vinegar and water were added and the cabbage slowly fermented turning into sauerkraut. We ate sauerkraut all winter.

Digging potatoes was also a Saturday job as it required lots of help. As soon as I was old enough to pick up a potato and put it in a bucket I was old enough to help. Dad used the team and a walking plow to plow along the potato row. The potatoes rolled out and most were on the top of the soil. We picked them up in buckets, and when the bucket was full we dumped them in the wagon. When all the rows were plowed, Dad hooked the team to the harrow and dragged the patch. This uncovered the potatoes that were under the dirt, and we picked them up.

After dragging the patch several times we had all the potatoes loaded on the wagon. The team pulled the wagon to the cave. We put the potatoes back in buckets and carried them down into the cave for storage.

In late winter we spent another Saturday sitting in the cave picking up each potato and breaking off the sprouts. We also discarded any that were starting to decay. The cave, located just northeast of the kitchen door, was a good place to store vegetables as the average winter temperature was around 50 degrees. It never froze. We stored all the canned goods, potatoes, carrots, beets, parsnips, and squash in the cave. We also kept the eggs and cream there for the creamery truck.

The yield of garden produce varied greatly from year to year. We were totally dependent on nature as there were no insecticides, pesticides, or irrigation to help maintain yields. During dry years there were grasshoppers that ate all the leaves off the plants. This really reduced the yield of vegetables.

12

— · —

4-H

Me with Gibraltar—my first 4-H calf

The 4-H club was one of the few out-of-school activities available for farm kids. This program was started in 1902 by a schoolteacher in Iowa to help boys become better farmers. Later it became a national program as part of the Federal Agriculture Extension Service. The Agriculture Extension Service was connected to the land grant university system in each state. The purpose of the program was to carry the results of agriculture experiments to the farmers.

A county agriculture agent was assigned to each county in Iowa. They held meetings, farm demonstrations, and visited farmers to improve farming practices. One method of teaching the farmers was to get their sons involved with agriculture. The county agent tried to have a boys' and a girls' 4-H club in every township. Adams County had twelve townships.

The boys' and girls' 4-H clubs were separate organizations. Many years later the clubs became co-ed but in my time it was thought the sexes should be separated as much as possible. About once a year there would be a county-wide rally and the boys and girls all attended.

To join a 4-H club, you had to be at least nine years old and live on a farm or an acreage where you could keep livestock. Some local farmer was selected to be the club leader and the club tried to meet once a month. In the summer we often had more than one meeting a month to make up for any missed meetings during the winter.

The members picked at least one project to have for the year. Most of the boys had beef steers, heifers, or pigs, but a few had sheep, rabbits, chickens, dairy heifers, or field crops. The big event of the year was the Adams County Fair in Corning.

Getting ready for the fair was a major activity in our house. The fair was usually held in July just before the grain threshing started. I always had beef steers to show, and they had to be tamed and broken to lead. I started by tying the calves to the manger during the day so they got used to the halter. In the cool of the evening, with Dad's help, I started leading them.

Effie Lee also had beef as a project, so she was in on the act. We worked every evening until we could lead the calves and make them stand evenly on all four feet so they looked good for the cattle judge. If the steers had horns, weights were placed on the ends of the horns to make the horn grow straight. Later there was a rule that all cattle needed to be dehorned so there wasn't the chance of injury to the exhibitor or other cattle from the horns. The calf's horns and hooves were polished and oiled so they shined. The hair was combed and the tail curled so the steer was attractive to the judge. We practiced doing this before we took the steers to the fairgrounds.

The day before the fair started, we got all our supplies ready. We had a big show box to store our combs and brushes. We needed a pitchfork to clean up behind the steers at the fair, several bales of straw to make a bed for the steers, sacks of ground feed, buckets to feed and water the steers in, and our show halters and show sticks. Dad didn't own a pickup (no farmer did at that time) or a trailer so we hired a truck to haul everything to the fair. We usually called another 4-H member and shared the truck with them.

The first day of the fair was really exciting—especially the first few years I belonged to 4-H. The truck arrived early in the morning, 7am, and we got everything loaded. The steers were tied to the sides with their halters and the supplies put in the back or on a platform above the steers. We followed the truck in our car to the fairgrounds.

After arriving at the fairgrounds we looked for our assigned stall in the beef barn. Our 4-H club members were all stalled together in one area. It was supposed to be the kid's project so I tried to lead my steers from the unloading area to the stall, but I was little and the steers were afraid of all the noise. I usually needed help to keep the steers from getting loose and running around the fairgrounds. Once we got the steers tied in their stall and everything in place we could look around at all the other steers and try to figure out who had the best steer.

At noon we met at the car and Mom had sandwiches and fruit ready for lunch. The fair was great fun with a merry-go-round, Ferris wheel, and other rides. In the afternoon there were horse races. In the evening there were different kinds of musical acts. We went home at five o'clock to do the chores, eat supper, and then returned to the fair.

Each day there was judging of different projects and open class exhibits. The 4-H classes were limited to 4-H members but anyone could enter in the open classes. In the early years there was a large number of draft and saddle horses judged, also purebred beef and dairy animals, and purebred swine. A large building held all of the girls' 4-H projects, including sewing, food preparation, and home decorations. The girls were allowed to show livestock in the same classes as the boys.

The big day had arrived: judging the beef steers. We did the morning chores early and arrived at the fairgrounds at 6:30am. The judging started promptly at 8am. We got the steers out and washed them, combed their hair, curled the bush of their tail, and tried to keep them clean until we were called for the show ring.

The steers were shown in classes by breed and weight. The judge, usually a man from the university, a county extension agent, or a noted beef breeder, looked at each animal to see if it matched the breed type, was at the proper weight and finish (fat over the muscle), and ready to be slaughtered. The steers were meat animals and were judged on meat qualities.

The judge lined the steers up with the best at one end of the line and the poorest at the other end. If there was more than one class of a breed, the top two animals of each class returned for the champion and reserve champion of the breed. The champion of each breed then competed for the grand champion steer of the show. I never had the grand champion at the county fair, but my brother Don and I think Elinor both had a grand champion.

To show at the state fair in Des Moines you had to be at least 12 years old. At that time the steers remained at the fairgrounds the entire duration of the fair. Today the 4-H steers are there for only two or three days.

We pitched a tent in the campground and stayed for a week. Dad went home for a few days to do work and we arranged for someone to do our chores while we were gone. The Iowa State Fair was great fun, and I roamed with friends around the fairgrounds and watched all the activities. We still had to check on the steers and keep them clean, fed, and exercised.

During my 4-H days I took steers to the St. Joseph, Missouri livestock show, the American Royal at Kansas City, and the Aksarben in Omaha. I had a good time, and I am glad my dad and mother were invested in this activity.

13

CHORES

*Corn sheller photo by Asp. courtesy of
Wikipedia—Creative Commons license
CC0*

Every farm kid had chores to do, and we all knew what it meant when someone said they had to get home to do chores.

Chores varied from farm to farm depending on what livestock a farmer raised, the time of year, and morning or night. Some chores, like milking the cows, were done twice a day rain or shine, every day of the week. Others, like taking the hay to the beef cows in winter could be done every other day and usually wasn't done on Sunday.

As a boy growing up on a farm I started with light chores and grew into the heavier work. Dad had a saying, "He was too heavy for light work and too light for heavy work." The chores I was assigned to do depended on what I could lift.

The first job I can remember was helping Mother pick up corn cobs in the hog lot. Even a three-year-old can pick up cobs if given the right directions. The cobs were carried to the house and used in the cook stove. We had a big wooden box that sat by the stove to store the cobs and other small pieces of wood to use in the fire.

In a few years I could carry a bucket to the hog lot and pick up cobs by myself. Dad fed the hogs ear corn, and they ate the kernels off and left the cobs scattered around the lot. When I showed up with the bucket the hogs were looking for something to eat and were a real problem unless I gave them some more corn. Filling the wood box with cobs and wood was an evening chore that was done two or three times a week.

Shelling corn for the chickens in the laying house and for some of the small calves was also an evening chore and needed to be done every night unless I shelled extra one evening for the next day's feeding. The corn sheller had a handle I turned until it was up to speed. There was a heavy flywheel that provided momentum that kept the sheller working. Without the flywheel, the sheller stopped when I dropped an ear of corn into it. The corn kernels were stripped off the ear as the ear of corn turned around teeth in the sheller. The cob traveled up the sheller and dropped out in the front. The corn kernels dropped out the bottom of the sheller into a bucket placed under the sheller.

I have mentioned the chores relative to raising chickens and milking cows. When I was older one of the most time-consuming chores was cutting ear corn into small round pieces to feed cattle. I used a corn knife which was like a machete. It was about two-feet-long and had a wooden handle. The blade tapered from two inches at the handle to three inches at the end. I held an ear of corn in my left hand and swung the knife with my right, cutting about a one-half-inch slice off the ear. I needed to have two or three bushels of cut corn so it took some time. The knife was sharp and one night it bounced off the ear of corn and hit my left thumb resulting in a big gash. I still have the scar.

Other regular chores were carrying the skimmed milk to the hogs and pumping water to fill the water tank for livestock. I took feed and water to the sows while they were in their pens with their little pigs. The horses were fed in their tie stalls in the barn. The stall had a box in the manger to put oats or corn and hay was put in the manger. The grain was usually fed first and the hay later.

When I was assigned a chore, I was expected to do it every night or morning without being told to do it again. In the winter it got dark early and I needed to carry a kerosene lantern to see what I was doing. There was a special hook on the wall behind the cows to hang the lantern. Dad warned me to be careful with the lantern in the barn so I wouldn't light the barn on fire.

When the milking and milk separating was finished all the chores were completed. In the winter we always finished the evening chores before we ate supper and it was usually 7pm before Dad and I were in the house. In the summer we ate before finishing the chores as there was plenty of daylight to finish them.

Originally published in Southern Planter, vol 66, May 1905, pg 428. Made available by Google. This work has passed into Public Domain.

Hauling Manure

A big job on a livestock farm was cleaning up after the animals: hauling manure. During the winter the animals were confined to buildings and sheds. The milk cows spent the night and most of the days inside, and the area had to be cleaned daily. When I was in high school I spent many Saturdays helping Dad clean out buildings including the chicken house. If the weather was nice the manure was loaded into the manure spreader and hauled to a field that day.

The manure spreader was built just for the purpose of hauling manure. It was a wagon that had a chain with slats that moved the load to the back. On the end of the wagon were round, fan-like beaters that threw the manure off the wagon as it was pulled across the field. The large rear wheels were the drive wheels. The spreader was pulled by two horses. A lever by the driver's seat was used to put the spreader in gear and control the speed of the apron.

In the winter it was important to make sure the spreader was completely clean after unloading. If any manure was left in overnight it froze the apron. This stopped the use of the spreader until the weather warmed enough to thaw it out.

Manure was the main source of fertilizer for the crops. The fields close to the farm building were the most productive as more manure was spread there. If a farmer was in a hurry, he unloaded the spreader and returned to do other chores. It took time and planning to drive to the far end of the fields to spread the manure.

The manure was mixed with the bedding used for the livestock. To load the spreader we used a fork with five tangs. A pitch fork with three tangs was used for haying and threshing.

When the spreader was loaded, I drove the team to the field and spread the manure. Dad was doing other chores while I was gone. We used the horses several years after Dad purchased a tractor, but eventually he shortened the tongue and pulled the spreader with the tractor.

Raising Oats

The first crop to be planted each spring was the small grain. Small grain crops included oats, barley, and spring wheat. Dad usually planted oats and a small amount of wheat. The fields seeded with the small grain had grown corn the previous year. The ground was prepared by using a disk pulled by four horses.

The disk was seven-feet-wide and a heavy load. The levers in front of the seat were used to control the pitch of the disk.

When the levers were pushed forward, the middle of the disk slid back and the blades cut into the soil. The disk blades cut from two- to four-inches-deep and rolled over the soil. This made a soft seed bed where the oats were planted.

I never drove the horses on the disk. I was too small to handle four horses and it was a dangerous piece of equipment. If you happened to fall forward off the seat you were in danger of serious injury from the disk blades.

The oats and clover or alfalfa seeds were seeded with an endgate seeder. This seeder fit on the back of a wagon and was driven with a chain that went around a cog wheel fastened to the rear wagon wheel. It was the same method as was used on the spreader. After the seeds were scattered the land was either harrowed or disced lightly to cover the seeds. Oat seeding was done in late March or early April depending on how early the fields warmed up. When harvested, the oats were used for horse and chickenfeed.

When the seeding was finished, the small grain required no additional care until the grain was ripe and ready to harvest. It was a pretty sight when the fields of grain turned a golden color.

In early or mid-July the small grain was ripe enough to cut. The cutting process needed to be done before the grain dried and the seeds shattered out of their heads and fell to the ground. The cutting was done by a machine, drawn by four horses, called a grain binder or reaper. This was a complex machine with lots of moving parts.

The machine was invented in 1834 independently by Obed Hussey in Ohio and Cyrus McCormick in Virginia. Cyrus McCormick started a factory in Chicago that became the International Harvest Company, a big farm equipment manufacturer during the late 1800s and the 1900s.

As the machine was pulled through the standing grain, a revolving reel would push the grain across the cutting sickle and lay it on a moving canvas. The canvas carried the grain into the binder. A knotting device invented in 1880 wrapped and tied each bundle of grain with twine. The bundles dropped on a wire carrier and when eight or ten bundles were on the carrier the operator dropped the carrier to the ground and the bundles slid off in a pile. The carrier looked like a big fork, and when it dropped to the ground it collapsed. The friction from the ground slid the bundles off.

The operator of the binder had to drive the horses, watch the sickle to be sure it was cutting properly, and make sure the twine was feeding through the needle and knotting machine. It was a busy job.

My only involvement was pulling the binder with the tractor after Dad purchased a tractor and quit using the horses.

Once cut and bundled, the grain bundles were stacked by hand in shocks of eight or 10 bundles. The shocks were made by standing two bundles butt down together and continuing to place bundles around until there were eight or 10 bundles together.

One or two were laid flat across the top. This allowed the grain to continue to dry until the threshers arrived. Fields of shocked grain were very pretty and were used on lots of ads in magazines during the 1940s.

14

THRESHING GRAIN

Threshing crew at the Quiggle farm north of Amboy, Minnesota. This file was contributed to Wikimedia Commons by Blue Earth County Historical Society as part of a cooperation project. The donation was facilitated by the Digital Public Library of America, via its partner Minnesota Digital Library. No Copyright - United States

I was eight years old and had a terrible time sleeping one August night in 1937. Why? The threshing crew was coming the next morning. What an exciting time for an eight-year-old boy!

Mother, Effie Lee, and I had been to the neighbors that day while they were threshing. It was like a small entertainment park with activities everywhere around the farmstead. The owner of the big threshing machine and steam engine that powered it was in charge of the crew and its operation. I was always amazed at the organization of the threshing crew.

During the summer before the small grain was ripe, the thresher contacted all the farmers in the area. He asked if they wanted to be part of the crew, how many acres of grain they had to thresh, and what job they wanted to do during the threshing season.

In some locations the crews were organized into a company and jointly own the thresher and steam engine. The company elected officers that ran the company and set up the threshing season.

The working crew consisted of approximately 20 to 25 men, eight bundle wagons, two or three grain wagons, a water wagon, and 16 to 20 horses. Two or three men were in charge of the steam engine and threshing machine. Water and fuel was hauled to the steam engine, and the threshing machine needed frequent greasing and a constant watch for breaking parts. One man ran the straw blower to even the straw stack or move the straw at the direction of the man on the stack. Two men hauled the grain to the grain storage place and scooped it into the granary. One of the men was left-handed so they could stand face to face and both scoop the grain.

In the field, six to eight bundle wagons hauled the bundles to the threshing machine. Four to six men loaded the bundle wagons by pitching the bundles to the man on the wagon; he placed the bundles to get a large load. It was always hot in August and the men working in the field needed water so there was always a water boy.

Freddie and I filled this position for two or three years. Dad purchased a small saddle that fit him and I was off to work. I carried two jugs of water that hung on the saddle horn. The jugs were the old ceramic type, and were wrapped with burlap. I filled the jugs at a well and soaked the burlap with water to help keep the water cool. There was no ice to use in that time.

A corn cob was used as a cork to keep the water from splashing out of the jug as Freddie trotted along. I went from wagon to wagon. The men pulled out the cob, splashed out some water to clean the jug opening, and took a drink. I was busy all day going from wagon to wagon and back to the well for more water.

I rode home on the bundle rack and led Freddie from the rack as I was tired of riding in the saddle all day.

The first crews to arrive in the morning were the bundle racks and the men to do the pitching. They headed out to the grain field and started loading the bundle racks. The big steam engine slowly moved down the road pulling the threshing machine. The location for the straw stack had been selected and the engine pulled the machine to the desired spot. The wind direction was a factor in placing the machine as the wind needed to be blowing from the front to the back so the dust and chaff blew away from the men unloading the bundle racks.

The machine was leveled by the operator digging holes for the wheels, and sometimes it took several tries until it was level. The steam engine was then unhooked from the thresher and was turned around to face it. A long belt was stretched from the engine to the thresher and placed around the drive pulleys. For some reason the belt was always twisted. I was never sure just why but think it had to do with the wind and the twist kept the belt on the pulleys. The belt was about 50-feet-long, and I think this helped in keeping the thresher running at a constant speed.

The machine was ready to run. The first loads of bundles arrived, and the threshing began. A bundle rack was on each side of the big hopper where the bundles, grain head first, were placed into the machine. The thresher had a knife that cut the binder twine and let the loose grain straw flow through the machine. Little hammer-like devices knocked the grain out of the heads.

The grain kernels dropped onto moving grates with holes that allowed the grain through. The straw continued moving to the back of the machine and a big fan blew the straw out onto the stack. At the bottom of the thresher, the grain collected and was augured up into a weigh bucket. When the bucket filled with grain, it dumped the grain and counted the number of bushels that were harvested. The farmer paid his threshing bill according to the number of bushels harvested.

The farm wives and daughters were also involved during the threshing season. Every morning they would arrive at the threshing farm with food. A lunch was provided to all the workers in midmorning. At noon there would be a large meal with all kinds of meats, vegetables, and of course, pies. Everything made was fresh from scratch.

Makeshift tables made with planks were set up in the shade of a tree. One table was used to spread out the food and one table was where the men ate. The thresher was shut down during the lunch period so everyone got to eat. Some men stretched out in the shade and took a short nap. The women spent the afternoon cleaning up and getting another lunch ready for the men around three o'clock. The thresher did not stop for the morning or afternoon lunches and the men ate on the run.

In the 1940s the big threshing rings stopped as many farmers were buying grain combines to harvest their grain. Dad found a small, used threshing machine that he could power with his tractor, so he and some neighbors continued to cut and bind the grain and run it through the thresher. They only used one bundle wagon, and they went to the field, loaded the wagon, brought it back to the thresher, and threshed the load. Then they went back to the field for another load of bundles. It took several days to get all the grain threshed but that was still less time than the month spent on the threshing ring.

The first combines were invented in 1913 and marketing to farmers started in 1915. These were horse-drawn machines and I don't remember ever seeing one in operation. The farm tractor was invented, and the first tractor-pulled combine was introduced in 1925. In 1942, the first self-propelled combine was introduced and that really changed grain harvesting.

I can remember how the old-time farmers were so upset with the combine. The threshing machine was always set perfectly level, so how could a machine run up and down hills and not spill all the grain? The difference was the way the working parts were designed to keep the grain contained on the slopes.

15

— • —

RAISING CORN

*This type of wagon was used on Iowa farms to pick corn, drive
to town to haul supplies, haul grain, and do other work around
the farm. It sold for $41.45 in the Sears catalog of 1908.*

Corn was the main crop for all livestock farms during the 1930s and 1940s
as corn was used as feed for animals. Today corn is raised by most farmers as a
cash crop. The only cash income for farmers then was from the sale of animals.

Getting the corn planted in the spring was a big job and was the source of a
lot of worry for Dad. Livestock grazed the land during the winter and ate most
of the leftover corn stalks. He used a plow to prepare the soil for planting, and
plowed the soil in the spring.

He used a four-horse team to pull the plow. It was a two-bottom plow with each plow cutting a twelve-inch furrow three- to four-inches-deep.

Levers were used to adjust the depth of the plow. Sharp plow shares sliced through the soil to lift and roll it over. The shares were sharpened at a blacksmith shop every few days. There was a blacksmith at Nodaway and one at Corning.

To plow a field, Dad started at the edge of the field and plowed around the field lifting the plow for the corners. The field was laid out long and narrow and as the plowed areas got closer together Dad skipped the ends. He then plowed the corners leaving a dead furrow down the middle and out to each corner.

Extra discing was done to fill the dead furrows with soil. A square forty-acre field would measure one mile around the outside. Most of our fields were not square due to the ditches and pasture areas, but I will use a square forty as an example to illustrate the time it took to do the spring plowing.

Horses pulled the plow at approximately three miles-an-hour. In one hour the team made three rounds, or six feet of plowed soil. This was equal to approximately 1.4 acres. To plow 80 acres of land it took nearly 60 hours of actual plowing time. This did not include the time it took to harness the horses, let them rest every hour, and get to and from the fields.

The last thing to do each day before putting the team in the barn was to hook them on to the harrow and harrow the fresh plowed soil. In the spring it was often wet and the plow turned up big slabs of the soil. If the slabs were not broken up, they dried into big hard chunks (clods), and it was difficult to get them smoothed out to make a nice seed bed for the corn.

I spent many Saturdays and time after school in the spring walking behind a harrow. One day Dad purchased a harrow buggy that was fastened to the harrow, and I could ride instead of walk. Walking all day in a plowed field behind a harrow sure helped one sleep well at night.

The field was plowed, harrowed, disced, harrowed, and now was ready to plant.

The corn was planted in May as the soil needed to warm to 55 degrees for the corn seed to sprout. Planting corn was a real art, and farmers judged each other on how straight the rows were in the field. I don't know if they ever spoke outright to each other, but they sure made comments to each other on how the neighbor's cornfield looked. Dad always used the best team (they walked the straightest and followed the mark) to pull the corn planter.

I was never allowed to plant corn as Dad wanted the rows straight. The corn planter was a sophisticated machine that had been developed over many years starting with the hoe. It was pulled by two horses and planted two rows at a time. The rows were usually 42 inches apart as this gave room for a horse to walk between the rows during cultivation and corn picking.

On the first trip across the field a check wire was unrolled. This wire had a knot every 42 inches. On the return trip the wire slid through the planter gear and tripped the planter to drop the seeds every 42 inches. The rows then ran both ways across the field. This allowed us to cross the field to cultivate out the weeds in the rows.

If the wire broke, there could be a gap and the cross rows were crooked. The person on the planter always carried a piece of wire to splice the wire if it broke. There was a marker on each side of the planter that dropped down and drug along the ground on the side where the team returned. Good teams soon learned to keep the mark between them and walk straight.

Planting corn was a slow process. A good team could plant from one to one-and-a-half acres an hour, depending on the length of rows and the time it took to turn on the ends and move the wire. A team worked for eight hours, so in a day a farmer planted from eight to 12 acres.

The corn seeds planted were saved from the crib and were called open pollinated. This was before the development of hybrid seed corn. One of the dry years (1934 or 1936) Dad was one of the few farmers that harvested any corn. As he scooped the corn off the harvest wagon, he picked out the best ears and saved them. He took a sample of four kernels from each ear, put a tag with a number on each ear, and placed the kernels on a square in a rag germination test.

The rag doll test was made with a square piece of cloth marked in two-inch squares, each having a different number. The cloth was rolled up and kept warm and moist. If the kernels germinated, the ear was saved for seed; if the kernels rotted, the ear was used for feed. The good ears were hand-shelled and the kernels sold to other farmers for seed corn. I think Dad sold seed corn several years.

The corn plants soon emerged from the ground, and I can remember how small and yellow the plants looked. At that time there was no fertilizer to add to the fields except the animal manure that had been spread on the fields. The weeds grew as fast as or faster than the corn and the only way to control the weeds was to cultivate as this was before any herbicides were used.

Dad had a one-row, horse-drawn cultivator. The cultivator had a seat that was over the corn row so I could ride and look down on the row. When the corn was little or if the ground was cloddy I had to watch to see if any of the corn plants got covered with dirt. If they did, I stopped the team and uncovered the plant.

Shields (flat metal strips) were used to keep the dirt off the plants. I had my feet on the cultivator shanks and pushed them apart if a corn plant was out of the row and to hold the shovels down in hard soil. After the first cultivation the shields were removed and the shovels set closer to the rows so the soil covered the weeds close to the corn plants.

I really enjoyed cultivating corn. The weather was warm, there were no flies bothering the horses, and they walked along without any problems. Every so often I got off and walked around to get the kink out of my neck from watching the corn. The horses must have had built-in clocks, because if you tried to start a new row at noon they just stood as they knew it was time for dinner. At five it was the same routine; they didn't move except toward the barn.

The corn plants grew slowly, and there was always worry about the weather. It was either too dry, too wet, or too cold for the plants. The popular measurement for corn was "knee high by the Fourth of July" to make a good yield. It took a good weather year for most of the fields to be knee-high by the Fourth in southern Iowa.

Today, corn in Iowa will be over your head and in tassel by the Fourth of July.

16

HARVESTING CORN

Photo of corn bassinet by Photoshot courtesy of Pixabay

Harvesting corn, commonly called "picking corn," or "husking corn," was an intensive manual labor enterprise. The corn was usually dry enough to start picking and filling the cribs in late-September or early October.

To open a field, the team walked into the field leaving three rows to the outside next to the fence. One person picked the first two rows, and if there were two people, Dad and I, we picked three rows. Once through the field the team turned around and walked back straddling the first row. The fields were usually laid out in lands of approximately 100 rows. This reduced the time it took to turn around at the end.

The corn was harvested by hand by pulling ears off the corn plant and throwing them into a wagon. The wagon box was 26-inches-deep and was fitted with a "bang board." The picker threw the ears of corn toward the wagon and the ears bounced off the board into the wagon.

The bang board was made of two twelve-inch boards, two eight- or ten-inch boards and two or three six-inch boards. The boards had cleats that slid down on the top of the wagon box and then fit on each other. The wide boards were on the bottom and the narrow ones on the top.

As the wagon was filled with the ears of corn, a board was moved from the bang board to the other side of the wagon box. When the picker first started there was a high board to hit with the corn ears. As the wagon filled, the board got shorter, plus the side next to the picker got higher. The wagon box would hold a bushel of corn per inch.

To have a 50-bushel load you needed 50 inches of wagon, 26 inches for the wagon box, and 24 inches of the bang boards that had been moved over. Really good corn pickers could pick over 120 bushels in a day. It took three wagon loads.

At the end of the wagon was an end gate. This was a solid board, 60-inches-high with tapered sides that fit around the wagon box. It was hinged at the bottom so it could swing down to a level position with the floor of the wagon box.

The corn was scooped out of the wagon and thrown into the corn crib. The corn was handled twice in the same day by the corn picker. I can remember many nights when I had most of the chores done and Dad was unloading the corn by the light of a lantern.

To pick corn there was some necessary equipment. Corn kernels and the shucks were hard on the hand so everyone wore a special glove. The gloves were made of flannel and had two thumbs. When one side of the glove was worn out, you could put the glove on the other hand and have a new glove. Dad bought the gloves in packages of 12 as they were only worn a day or two.

There were two tools a picker could use to open the shucks. One was a peg; it had a leather strap your fingers fit in and a sharp metal point that slit the corn shuck so you could grab the ear. The other was called a hook. It was a metal hook that was fastened on a leather strap that fit in the palm of your hand. You pulled the hook across the shuck and grabbed the ear, snapped it off the stalk, and threw it to the wagon—all in one motion.

You could hear the bang, bang, bang of the ears hitting the board all across the countryside when the corn harvesting was going on. Some farmers took great pride in being the first one out to the field in the morning.

Harvesting corn was a long process. Dad's goal was to have all the corn picked by Thanksgiving, but we didn't very often make it. The neighbors talked if corn was still in the field by Christmas, and everyone wanted to be finished before New Year's Day.

There were problems in getting the corn picked. When the days got shorter there was less time to pick as it was difficult to pick in the dark. Snow storms stopped the picking. If you tried to pick in the snow, your hands got wet and so cold you couldn't work, and of course no one wanted to pick corn in the rain. The weather often determined how quickly the corn was harvested.

I never really enjoyed picking corn. I often got a rash on my arms that itched like mad, plus I was always so little and I wasn't very good. Seventy-five bushels in a day was about my best. Getting the corn harvested was a high priority and it was an acceptable absence from school to be home to pick corn. I usually missed two days a week to help Dad pick corn.

The ear corn was stored in a slotted crib. The slots in the wall allowed air to flow though the corn and dry it out. If the moisture was too high (above 18 percent) the corn rotted in the crib. A lot of the corn was fed to hogs and cattle on the ear, some was shelled for the chickens with the small corn sheller, and some was ground for the hogs.

Several times during the winter a big corn sheller came to the farm and shelled several hundred bushels. The cribs were designed with narrow doors along the bottom of the crib wall. They could be opened and the ear corn would roll out of the crib. The corn sheller was mounted on the back of a truck and the truck motor provided the power for the sheller. It took several men to keep up with the sheller. It was one of the many jobs on the farm where neighbors helped out. Sometimes the sheller was at two farms in a day, but usually it was an all-day job to shell and store the shelled corn in a bin.

Planting corn changed dramatically with the adoption of the tractor and tractor-pulled implements. Today a big tractor can pull planters with up to 24 rows. Row width has narrowed to 15 inches or less, and corn seeds are drilled being placed eight to 10 inches apart. Harvesting is also done by big self-propelled combines that shell the corn as they pass through the field. A big machine can harvest more than a thousand bushels in a day.

With the use of hybrid corn seed, fertilizer, insecticides, and herbicides, a corn producer can harvest more than 200 bushels per acre. This is a big change from the 35 bushels per acre that was harvested in the 1930s.

17

HARVESTING HAY

Harvesting hay was a major summer job on all livestock farms when I was a boy. The beef cows, dairy cows, and horses all depended on dried hay as their food supply during the winter months.

Hay was harvested from alfalfa and clover plants. The seeds were sown at the time of oat seeding with the endgate seeder. There was a separate box on the seeder for the seed, and if Dad was seeding on a Saturday, when I was old enough I rode with him and either drove the team or put the grass seed in the box. The seed was metered out and fell on the fans that threw the seed out on the ground.

The first year, the alfalfa or clover was seeded with the oats and was an established plant under the oat plants. The next year the plants were big enough to be harvested as hay. Alfalfa would last as a crop for several years while the clover only grew for the two years. The weather was an important factor in deciding when the hay cutting started. In those days there was no TV weather channel or Doppler weather satellite to forecast the next week's weather. Dad listened to the radio, if the batteries were up, and tried to figure out when rain was forecast.

Rain was bad for mowed hay. If the hay was nearly dry and it rained it caused the leaves to fall off the stems. The leaves contained most of the nutrition for the livestock. Farmers really bragged if they got all their hay in the barn without having it rained on.

The cutting was done with a horse-drawn mowing machine. The mower had two wheels that were the drive wheels, a tongue that went between the two horses, and a sickle bar. The sickle bar was approximately five-feet-long and folded up for transporting to the field. The cutting bar was set two or three inches above the ground.

It rode on shoes that were adjusted to rise or lower the bar. The sickle blades were riveted on the bar, and the operator had to carry new blades and rivets in the tool box to replace any blades that broke if you hit a stone. Guards were also placed on the bar. The guards had a sharp blade that helped do the cutting. The guards would also help in keeping rocks or sticks from hitting the blades.

The sickle bar was on the right side of the mower. To open the hay field for cutting, the first round was done with the sickle bar to the outside of the field. When mowing the first round, I had to be careful not to catch the fence. Once around the field I turned and went the opposite way and continued cutting until the field was finished. Some years, rabbits by the hundreds ran out of the field as the standing hay got smaller.

We always used the best horses on the mower. If you happened to cut through a bumblebee nest, the bees attacked the horses and you sure didn't want a team that would run away. I spent many summer days when I was high school-age mowing hay.

After the hay was cut, it took one or two days, depending on the weather, for the hay to dry. If the hay got too dry, the leaves fell off. If it was put in the barn too green there was always a chance of having a fire and burning the barn down. When the hay was just right, the field was raked using a dump rake. The rake was pulled by a team of horses, and the driver pushed a lever that caused the rake to lift and drop the hay.

The first job was to rake the hay into a long row, called the windrow. The next job was to rake the windrow into piles. When the hay was loaded onto a hay rack, one man walked along the row and pitched the piles of hay up to the man on the rack. He loaded the hay by placing it around the rack.

The load of hay was taken to the barn where it was stored for the winter feeding of livestock. The hay rack was unloaded into the barn using a big hay fork that lifted the hay. A hay rope was strung through a pulley system that allowed the rope to drop down with the hay fork to the hay rack.

The rope went along the top of the barn and down to the ground where it was hooked to a horse. A large door at the top of the barn dropped down to provide a place for the hay to enter the hayloft. The man on the load of hay set the fork and yelled for the horse to start to pull up the hay. Leading or riding the horse was one of my early jobs in helping with the haying.

The fork full of hay was pulled up to the top of the barn where it would attach to a carrier on a track that ran the length of the barn. When the fork reached the desired location, the man working in the hay mow yelled. The man on the load would pull a rope to trigger the fork and the hay dropped. The man in the mow pitched the hay around to keep a nice flat surface and fill all the mow.

Putting up hay was a hot and tiring job and took most of the summer. In 1945 Dad purchased a hay loader that was pulled behind the hay rack by the team. The team straddled the hay windrow, and the loader pushed the hay up and out onto the wagon. This made haying easier as the hay windrows didn't need to be bunched, and no one needed to pitch the hay onto the wagon.

In the 1950s the hay baler was invented and most farmers were soon baling their hay and storing it in bales. The first balers were stationary, and the hay or straw was hauled to the baler for baling. The next step was the addition of a hay pick up attachment to the front of the baler. This fed the hay into the baler. A man or boy rode on each side of the baler and fed the wires into the hay and tied it making a bale. This was a dirty job. Thank heavens Dad never owned a baler like this or I might have had to punch the wires in the dirt.

The dump rake was replaced with a side delivery rake that made nice little windrows that could be picked up and baled. Today most of the hay is baled in large square or big round bales.

18

— • —

BUTCHERING

O ne of the big advantages to living on a livestock farm was that we always had plenty of meat to eat. In the summer we had farm-raised chickens and in the winter we had fresh pork and beef.

One problem was storing fresh meat very long without it spoiling. Two methods used to store meat were canning and salting. The fresh, butchered beef was cut into the normal cuts and then cut into cubes that were placed in the glass jars and canned. The canned beef was available for use for several months after canning.

Pork was stored by using a salt mixture to cure the meat. Both a dry salt and a liquid that was injected into the meat were used. The meat was wrapped in waxed brown paper and tied with a wire. Dad hung the meat in the milk house with a wire so no rat or mouse could get in it. The sides of the pork made the bacon and the legs were cured as ham. The process today uses some of the same material to cure the meat.

The second problem with farm-raised meat was the amount that needed to be consumed. Hogs weighed nearly 200 pounds when they were slaughtered, and even a small steer was 800 pounds. This was a lot of meat for a family of four. To help solve this problem, neighbors got together. They butchered a hog at one farm, and the next time butchered a hog at a different farm. The families shared the meat. Dad also sold quarters of beef to different merchants in Villisca and this helped deplete the supply.

It was December 15, 1935, and it had been below freezing for two days. The radio weather reported more cold weather. Dad decided it was a good time to butcher some hogs. He talked to a neighbor up the road who was willing to help and take a hog for his family.

A 50-gallon steel barrel was set on a rack under a strong limb of the big elm tree. The barrel was filled with water and a fire started under the barrel with corn cobs and wood. It took over an hour to get the water boiling. A block and tackle was hung to the elm tree limb. This was used to dip the hog carcass into the boiling water.

A hog was selected in the hog lot and driven or pulled close to the barrel. The hog was shot in the head, and its jugular vein was cut with a sharp knife. The hog was ready to butcher. A single tree (a single tree is what a single horse's tugs were fastened to for pulling a wagon) was hooked into each of the hind legs and fastened to the rope on the block and tackle.

The men hoisted the hog up over the barrel and lowered it into the hot water. It was raised and lowered several times to make sure the hair was scalded. The hog was then lowered onto a table made with sawhorses and an old door. The hair was scraped off the hog with round, bell-shaped scrapers. As soon as all the hair was removed, the hog was pulled up again, split down the belly, and the intestines were removed.

The heart and liver were saved in a clean bucket and taken to the house. The carcass was split and the two sides were hung from the tree to cool. After the evening chores were finished, Dad cut the hog into quarters and carried a quarter at a time to the basement. The neighbor cut his hog into quarters and took them home in his car.

Dad used a meat saw, cleaver, sharp knives, and a hand-cranked grinder to process the hog. The head had been removed when he split the carcass, and Dad never processed any of the head. Some people made head cheese and also used the intestines for the sausage casings, but Dad didn't.

The first thing Dad did was to cut off the lower legs, then trim out the hams and the bacon slabs. We used fresh pork for a few days but most of the bacon and hams were cured with the salt. The skin was left on the hams and bacon to hold the meat in place and aid with the curing. The back fat, up to two inches, was trimmed off the loins.

The fat was used to make lye soap that Mother used to wash clothing. Small pieces of lean meat were ground and seasoned into sausage. Some of the loin that wasn't used for pork chops was cut into two-inch squares and canned in glass jars. The meat was cut into small portions, enough for one or two meals, and wrapped in a brown paper that was waxed on one side. In the winter, fresh meat was stored in a container outside where it stayed frozen.

Butchering a beef followed the same procedure except the beef carcass was skinned and not scalded as were the hogs. There was a lot more beef to process and Mother usually canned a hundred jars of beef during the winter. Canned beef and mashed potatoes and gravy was a favorite dinner of mine. The beef was processed in the basement using the same equipment as used for the pork.

Dad made extra income for the farm by selling quarters of beef to merchants or families without beef cows. Sometime in the early 1940s a locker plant opened in Villisca. We could haul a steer or hog to the plant and they butchered and cutup the carcass. The meat was wrapped, frozen, and stored in a small cabinet the folks rented from the locker owners.

Every Saturday night the last stop on the way out of town was the locker plant to pick up some meat—and ice for our ice cream.

19

HOLIDAYS

We didn't celebrate many holidays when I was little, and of course the chores had to be done even on holidays.

Christmas was the most exciting holiday as there was always the expectation of some great gift, like a pony or a fire truck. There was always a program at church on Christmas Eve. The family tradition was to do the chores early and have oyster soup before going to church. This tradition of oyster soup continued as long as Mother and Dad were alive and able to entertain the family.

We didn't have a fireplace, so we hung our stockings on the backs of chairs in the dining room. Santa came during the night, and we ran down the stairs early in the morning to see what he had left for us. Dad kept us in bed until he had the furnace going and the house was warming up. This was the only morning he didn't yell up the stairs to get us out of bed.

Each of us would have several gifts but they were usually socks, underwear, or something that was needed for school. Dad always threatened that if we weren't good, Santa would leave us a lump of coal.

At noon on Christmas Day we usually had a family dinner or went to someone's place for dinner. Grandmother Brown and some of Mom's sisters were usually there. I do not recall many gifts being exchanged. If there were any gifts, they were something Mother had made or some meat for those that lived in town.

Every year I always felt Christmas was so slow in arriving and over so fast. I guess it is still that way.

I do recall one special day. On December 24, 1936, we went to Uncle John's to visit for Christmas Eve. John, his wife Maxine, Aunt Ethel, and my family were there. After eating supper, Effie Lee, Don, and I were left in the kitchen with Mother. Ellie was just a baby and was probably sleeping.

The rest of the grownups were in the living room doing something. Suddenly, there was a rap on the kitchen door and Mother took us into the living room. There was no light on, but we could see a small train running around a track with sparks flying. The toy train was a wind-up train and became one of my favorite toys. I often wonder what eventually happened to it, but it was probably discarded when my folks moved or when Mother did some deep-cleaning.

From left, Grandma Jenny with Flossie
(Mom) and her sisters Lois and Louise

We always celebrated everyone's birthday. I don't remember having a party, other than with relatives, until I was in high school, but Mom always baked a cake and decorated it with candles. Of course we sang "Happy Birthday."

When I was little, I played with toys just like young people do today. The only difference was that I had to use things found on the farm instead of purchased toys.

My favorite game was to play farming. Farming was the only activity that I knew and it was easy to set up a game. All I needed was some string to make fences, acorns for pigs, small corn cobs for cattle, and I was ready to go. I spent hours adding to the farm, getting more animals, and in general having a good time. It was like the video games of today except I had to imagine the whole thing. We did have some purchased toys like little trucks and tractors.

Card games were great for families and in the evenings we often played cards, usually pitch. Chinese checkers was a big hit and neighbors would get together to play. The highlight of the evening was when a player could move a marble from the point of the star all the way across the board to the opposite star point. Later another card game, canasta, swept the country and people joined clubs to play. Checkers was also a popular game.

Memorial Day was a special day when I was young. The folks always took flowers to the cemeteries where relatives were buried. All the Narigons were buried in Nodaway and the Days and Browns were buried in Villisca. Peonies were usually blooming by Memorial Day, but if the spring was cold and there were no peonies other flowers were located to decorate the graves. I can remember picking wild bluebells, irises, and even taking apple blossoms to the grave sites.

Memorial Day was to honor the military personnel that had died in war, and a ceremony was held in one of the churches in Nodaway. "Taps" was played as the names of the deceased were read. A snare drum roll sounded after each name. When all the names were read, someone gave a speech and the surviving veterans, if able, marched in uniform the two miles out to the cemetery. Flags were placed on the veterans' graves and the ceremony was concluded. I think this traditional way of commemorating Memorial Day stopped at the start of World War II.

Thanksgiving was never really celebrated as a holiday on the farm. The only thing that was different was the mail wasn't delivered and school was out. Dad would still be harvesting corn and if the weather was good we would be in the cornfield picking corn. One year when I was in high school all the members of the Methodist Youth Class that were available came out and picked corn. It was a money-raising project for some activity. That was the most exciting Thanksgiving that I remember. Mother did usually fix a special dinner or supper so we could say we had celebrated.

For several years the merchants in Corning and Villisca had special bargains on the Saturday before Thanksgiving to start the Christmas season. There were drawings for live turkeys and one year in Corning live turkeys were tossed off the top of the stores for people to catch. I think this was done only once as there were fights over who had caught the turkey and some of the turkeys were hit by cars. After that the merchants decided it wasn't a good idea.

20

MY BEST FRIEND

When I was eight or nine, I started walking to the mailbox (one-quarter of a mile) by myself to get the mail. I soon became acquainted with Rufus Ritnour, an old man that lived with his wife a short distance from our mailbox. He was a real character and soon became one of my closest friends.

They lived on an 80-acre farm that was mostly pasture land. They milked four or five cows, and during the summer they milked outside. The cows just stood still in the barn lot, and they sat down and milked. If the cow moved they would follow her until she stood still again. I thought this was really something as we always milked in the barn with the cows in the stanchions.

They also raised a few pigs and Ruff purchased a young male hog from a neighbor and raised it for his boar. He drove a Ford Model A and hauled the pig in the back seat. The pig's legs were tied together to keep it from running around. On one trip the pig got his legs untied and tried to jump up into the front seat with Ruff. Ruff tried to knock it back and ran off the road into the ditch.

"By Grabs that dang pig almost got me killed," was his quote.

They raised a big vegetable garden and had a small orchard of peach and apple trees. Mrs. Ritnour, I can't remember her first name, made the best homemade pies, and if I planned my trip at the correct time I was always offered a piece of pie.

Ruff also kept honey bees and helped me get started with honey bees. He had his own method of building hives and supers (boxes for the honey) to place on the hives. He used a 1x12-inch board for the sides, ends, and bottom of the hive. On the top he placed laths with spaces just wide enough for the bees to crawl through. The supers were made from 1x6 or 1x8 boards, again with laths on the bottom and top so the bees could climb through them. There were no combs to place in the boxes and the bees just made the combs of honey any direction they wanted.

When they harvested the honey, they took out all the combs and melted them in warm water. The wax would float to the top and the honey was poured out into jars. Some of the honey was eaten as comb honey, but it wasn't in nice squares.

Ruff taught me how to build the hives and how to catch a swarm of bees. When I was in high school, I had several hives of bees and even purchased bees from a mail order house. The people at the post office were always upset when the bees arrived.

One year I was stung on my forehead right between my eyes. Both of my eyes swelled shut and I could not see for several days.

Ruff loved boxing. I think he boxed when he was young and always wanted us to come to his place to listen to the world heavyweight champion fights on the radio. One night when Joe Louis was in a championship fight, Dad, Don, and I were a little late getting there and the fight was over. It had lasted less than a minute.

I was like "Dennis the Menace" to Ruff, but it was a great part of my growing up. Listening to all the funny stories he told, I didn't need a radio or TV.

21

AUTOMOBILES

Photo of Model A Ford by peteyp8 courtesy of Pixabay

The first car I can remember that we had was a Model A Ford. I don't know what year Dad purchased it, but it was a four-door, brown in color, and ran very well.

The cars then had no radio, air conditioning, heater, or power windows. There were no turning signal lights, and when you made a turn, if anyone was behind you, you made a hand signal. You had to make sure the driver's side window was rolled down, by hand, and if turning right you extended your left hand and bent your arm up. If turning left you pointed straight out to the left. You bent the arm down if slowing or stopping. The traffic wasn't very heavy so the lack of signal lights wasn't a problem.

The car had leaf springs just like a buggy, and of course the roads were bumpy. It was a rough ride if the speed was over 35 miles per hour. Flat tires were a real problem as there were rocks in the road and nails around the farm yard. If you drove a car, you soon learned how to change a tire.

We never drove very far in the Model A. It was nine miles to Villisca, and 12 miles to Corning. These were the major trips we made. I expect there were some longer trips but I cannot remember any. We never took vacations as a family. One year Dad and Mother, and I think Aunt Ethel, went to the World's Fair in Chicago. I don't know who owned the car they drove but Effie Lee and I stayed with Grandmother Brown in Villisca.

In 1940 we went to Sidney, Iowa, 62 miles away, to the rodeo. We were surprised to hear an announcement come over the loud speaker for Virgil Narigon to report to the ticket office. We wondered who wanted Dad. It was a car salesman from the Ford dealer in Villisca that had been to our farm. He had decided to accept the price Dad had offered for a new car. Dad kept the old Model A and traded two horses, some cows, and money for the new car.

We still lived on a dirt road but more and more roads were graveled. With bigger road graders, the roads were smoother. The summer I graduated from high school, 1946, Dad, Mother, Don, and Elinor took a long trip to the west, while I stayed home and did the chores.

I purchased my first new car in 1953 and paid $1,600 for it. When the price of new cars went over $10,000, it was said that anything costing more than $10,000 should have a basement. How times have changed!

22

THE END

I hope you have enjoyed this story of my early life growing up on an Iowa farm. I guess as we get older, we have time to reflect on our lives and the changes we have seen. I think my Grandmother Brown (1879-1969) witnessed the most dramatic changes in the development of the world. She was born in a sod house with no electricity, heating, plumbing, or windows and lived to watch on television as a man walked on the moon.

A few of the United States statistics for 1903 follow:

- The average life expectancy was 47
- Only 14 percent of the homes had a bathtub

- Only 8 percent of the homes had a telephone
- A three-minute call from Denver to New York cost $11
- There were only 8,000 cars and 144 miles of paved roads
- Only 6 percent of all Americans graduated from high school
- 95 percent of all births took place at home
- Population of Las Vegas, Nevada was 30
- The average wage was $0.22 per hour
- Sugar cost $0.04 per pound. Eggs were $0.14 per dozen
- There were no airports or airplanes
- There were no credit cards
- Most women washed their hair only once a month using borax or egg yolks as a shampoo

We had an old family joke that goes when an old timer was asked if he had seen a lot of changes in his life, he remarked, "Yes, and I was against them all." What will the next 100 years bring?

23

ABOUT THE AUTHOR

Joe Narigon left the farm in the fall of 1946 and enrolled at Iowa State College (now Iowa State University) at Ames, Iowa. He studied Animal Husbandry (now Animal Science) and graduated with a BS degree in December of 1950.

On December 20, 1950 he and Betty Ann Rock of Griswold, Iowa were married in the Griswold Methodist Church. They moved to Marshalltown, Iowa where Joe worked for Iowa State Extension Service as a 4-H Youth Leader.

In April of 1951 he was drafted into the U. S. Army and after basic training in Fort Knox, Kentucky, Joe was assigned duty as an Army meat inspector and spent two years in the Chicago, Illinois area. He was discharged from the Army in May of 1953 and farmed with his dad and brother Don at the home place until the fall of 1954.

In December of 1954 Joe and Betty moved to Atlantic, Iowa where Joe was the farm manager of the Walnut Grove Research Farm. Four children were born to the Narigons; Edward, Thomas, Michael, and Nancy.

In 1968, Joe resigned from his position with Walnut Grove and accepted the position of Webster County Extension Director in Fort Dodge, Iowa. He continued working in Webster County until 1980 when he accepted a position as director of the Warren County Extension Service, located in Indianola, Iowa.

In 1984, Joe and Betty were divorced. Betty continued to live in Fort Dodge and worked at Friendship Haven Retirement Community. In 1990, Joe married Frances Phillips. The couple retired in 1994 and moved to Holiday Island, Arkansas. In 1999 Frances died after a brief illness of cancer.

December 20, 2000, Joe and Betty were remarried on the 50th anniversary of their first wedding. They moved to The Village Retirement Community in

Indianola. Betty passed away on Saturday, December 16, 2017 at the Western Home Communities-Martin Center in Cedar Falls.

Joseph Edward Narigon passed away at age 93 on Wednesday, March 1, 2023, at CedarStone Senior Living in Cedar Falls. He was preceded in death by his parents, wife, two sisters, a brother, and his son, Thomas. He was the last of his generation.

Joe is survived by his sons, Edward (Sara) of Cedar Falls and Michael (Kim Lewis) Narigon of Manhattan Beach, CA; daughter, Nancy (Warren) Green of Kansas City, MO; daughter-in-law, Nancy Narigon of Minneapolis, MN; 10 grandchildren and 11 great-grandchildren, all of whom he loved.